THERMODYNAMICS and HEAT ENGINES

Rahul Basu
Emeritus Professor UGC, JNTU
Professor at VTU

Principal and HOD, Visiting Professor at Various Colleges

Bangalore India

Table of Contents

THERMODYNAMICS and HEAT ENGINES by Rahul Basu

Nomenclature and Symbols

Thermodynamic Properties

- P = Pressure (Pa)
- T = Temperature (K)
- V = Volume (m^3)
- v = Specific Volume (m^3/kg)
- m = Mass (kg)
- U = Internal Energy (kJ)
- H = Enthalpy (kJ)
- S = Entropy (kJ/K)
- Q = Heat Transfer (kJ)
- W = Work Transfer (kJ)

Specific Heats

- Cp = Specific Heat at Constant Pressure
- Cv = Specific Heat at Constant Volume
- γ=Cp/Cv

Other Symbols

- ρ = Density
- μ = Dynamic Viscosity
- η = Efficiency
- $\dot{m}$ = Mass Flow Rate

PREFACE

Thermodynamics forms the foundation of all energy conversion processes and is one of the most important subjects in Mechanical Engineering. A sound understanding of thermodynamic principles is essential for the study of power plants, internal combustion engines, refrigeration and air conditioning, gas turbines, compressors, and modern energy systems.

This book has been written primarily for undergraduate engineering students and is aligned with the syllabi of major Indian universities. Emphasis has been placed on clarity of presentation, worked examples, examination-oriented explanations, and extensive question banks.

The text integrates fundamental concepts with practical engineering applications and provides a bridge between theory and industrial practice.

The author hopes that this book will serve both as a classroom text and as a useful reference for practicing engineers.

Rahul Basu
Bangalore

ABOUT THE AUTHOR

Prof. Rahul Basu is an Emeritus Professor associated with several engineering institutions in India. He has served as Professor, Head of Department, Principal, Dean, and Visiting Faculty at various universities and colleges.

His areas of interest include:

- Thermodynamics
- Heat Transfer
- Fluid Mechanics
- Computational Fluid Dynamics
- Finite Element Methods
- Materials Engineering
- Robotics and Artificial Intelligence
- Engineering Management

He has authored numerous textbooks, research papers, conference publications, and technical monographs.

Prof. Basu has extensive experience in engineering education, curriculum development, and research guidance.

HOW TO USE THIS BOOK

The material has been arranged in a logical progression from basic thermodynamic concepts to advanced engineering applications.

Suggested Study Sequence

1. UNIT I – Introductory Concepts
2. UNIT II – First Law of Thermodynamics
3. UNIT III – Second Law of Thermodynamics
4. Power and Refrigeration Cycles
5. Internal Combustion Engines
6. Compressors and Auxiliary Systems
7. Question Bank and Model Papers

Examination Preparation

- Study solved examples carefully.
- Practice numerical problems independently.
- Review chapter summaries before examinations.
- Solve previous university question papers.
- Attempt the question bank after completing each unit.

ENGINEERING CONSTANTS AND CONVERSIONS

Fundamental Constants

- Universal Gas Constant, R = 8.314 kJ/kmol-K
- Standard Atmospheric Pressure = 101.325 kPa
- Standard Gravity, g = 9.81 m/s^2

Temperature Conversion

- K = °C + 273.15
- °C = K − 273.15
- °F = 1.8°C + 32
- °R = °F + 459.67

Pressure Conversion

- 1 atm = 101.325 kPa
- 1 bar = 100 kPa
- 1 MPa = 1000 kPa
- 1 N/mm^2 = 1 MPa

Energy Conversion

- 1 kJ = 1000 J
- 1 kcal = 4.187 kJ
- 1 kWh = 3600 kJ

COMMON ABBREVIATIONS

Abbreviation	Meaning
IC Engine	Internal Combustion Engine
SI Engine	Spark Ignition Engine
CI Engine	Compression Ignition Engine
PMM1	Perpetual Motion Machine of First Kind
PMM2	Perpetual Motion Machine of Second Kind
SFEE	Steady Flow Energy Equation
COP	Coefficient of Performance
TDC	Top Dead Centre
BDC	Bottom Dead Centre
MEP	Mean Effective Pressure
RPM	Revolutions Per Minute
BHP	Brake Horse Power
IHP	Indicated Horse Power
FAD	Free Air Delivery

UNIT I

IntroductoryConcepts.

DefinitionofTHERMODYNAMICS:

- Thermodynamics is the science that includes the study of energy transformations and of the relationships among the physical properties of the substances which are affected by these transformations.

- The term 'Thermodynamics' was first used in a publication by Lord Kelvin in 1849. The first thermodynamics textbook was written in 1859 by William Rankine, a professor at the Universityof Glasgow.

- **ScopeofThermodynamics**

Although aspects of thermodynamics have been studied since ancient times, the formal study of thermodynamics began in the early nineteenth century through consideration of thecapacity of hot objects to produce work. Today the scope is much larger. Thermodynamics now provides essential concepts and methods for addressing critical twenty-first-century issues, such as using fossil fuelsmore effectively, fostering renewable energy technologies, and developing more fuel-efficient means of transportation. Also critical are the related issues of greenhouse gas emissions and air and water pollution. Thermodynamics is both a branch of science and an engineering specialty. Engineers are generally interested in studying systems and how they interact with their surroundings. To facilitate this, thermodynamics has been extended to thestudyofsystems through whichmatterflows, including bioengineering and biomedical systems.

AreasofApplicationofEngineeringThermodynamics

1. Aircraftandrocketpropulsion,
2. Alternativeenergysystems
3. Fuelcells
4. Geothermalsystems
5. Windturbines
6. Automobile engines
7. Bioengineeringapplications
8. Biomedicalapplications
9. Combustion systems
10. Compressors,pumps
11. Steamandgas turbines
12. Powerproduction
13. Propulsion
14. Magnetohydrodynamic(MHD)converters
15. Oceanthermal,wave,andtidalpowergeneration
16. Solar-activatedheating,cooling,andpowergeneration
17. Thermoelectricandthermionicdevices

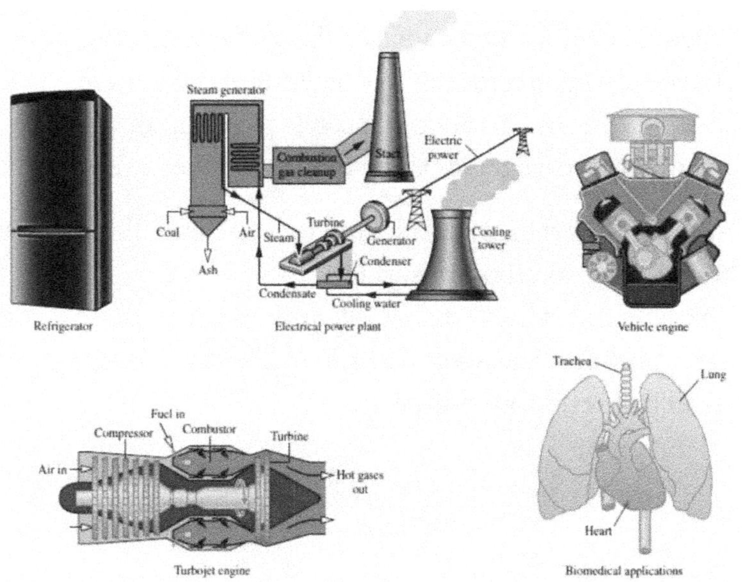

18. Coolingofelectronicequipment
19. Cryogenicsystems,gasseparation,andliquefaction
20. Fossilandnuclear-fuelledpowerstations
21. Heating,ventilating,andair-conditioningsystems
22. Absorptionrefrigerationandheatpumps
23. Vapour-compressionrefrigerationandheatpumps

ThermodynamicPropertiesandenergyrelationshipscanbestudiedbytwomethods.
(i) ClassicalThermodynamics
(ii) StatisticalThermodynamics

i. Classical and statistical Thermodynamics: It Involve studies which are undertaken without recourse to the nature of the Individual Particles which make up a substance and to their reactions. This is macroscopic view towards matter, and it requires no hypothesis about the detailed structure of matter on the atomic scale.
For **example**, the pressure of a gas in a container is the result of momentum transfer between the molecules and the walls of the container. However, one does not need to know the behaviour of the gas particles to determine the pressure in the container. It would be sufficient to attach a pressure gage to the container.
classical thermodynamics allows important aspects of system behaviour to be evaluated from observations of the overall system. It provides a direct and easy way to the solution of engineering problems.

ii. **Statistical Thermodynamics:** It is based on the statistical behaviour of large groups of Individual particles. This is microscopic view point of matter.
The statistical thermodynamics or microscopic approach to thermodynamics, is concerned directly with the structure of matter. The objective of statistical thermodynamics is to characterize by statistical means the average behaviour of the particles making up a system of interest and relatethis information to the observed macroscopic behaviour of the system.
For applications involving lasers, plasmas, high speed gas flows, chemical kinetics, very low temperatures (cryogenics), and others, the methods of statistical thermodynamics are essential.

System: A thermodynamic system is a Three dimensional region of space or an amount of matter, bounded by an arbitrary surface.
Everything external to the system is considered to be part of the system's **surroundings**.Thesystemisdistinguishedfromitssurroundingsbyaspecified b**oundary**, which may be at rest or in motion.

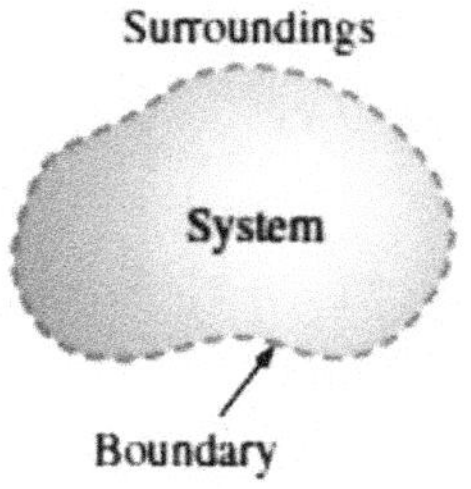

Systems may be considered to be *closed* or *open,* depending on whether a fixed mass or a fixed volume in space is chosen for study. A **closed system** (also known as a **control mass** or just *system* when the contextmakes it clear) consists of a fixed amount of mass, and no mass can cross itsboundary.Thatis,no masscanenterorleaveaclosedsystem,asshown infig. Howeverenergy may cross the system boundary.

Examples:

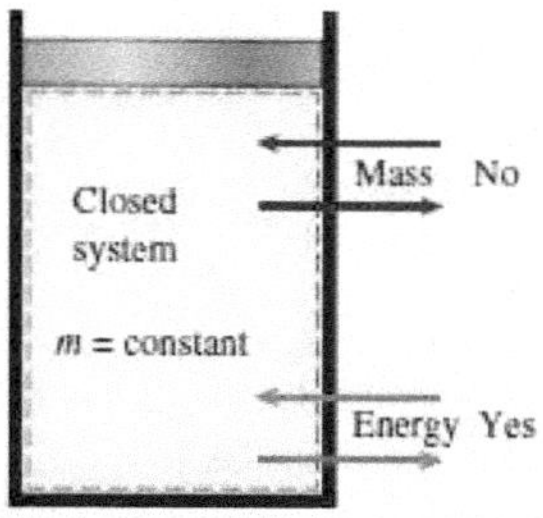

Mass cannot cross the boundaries of a closed system, but energy can.

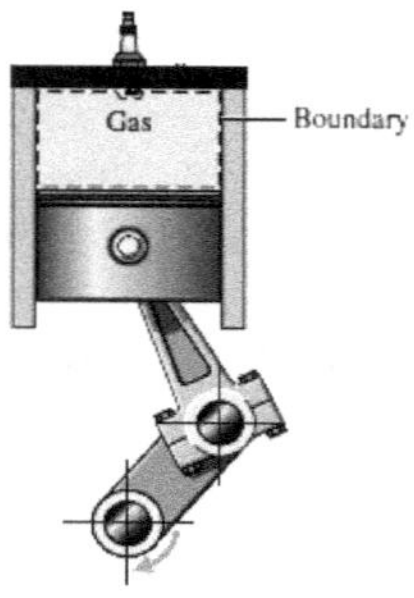

Closed system: A gas in a piston–cylinder assembly.

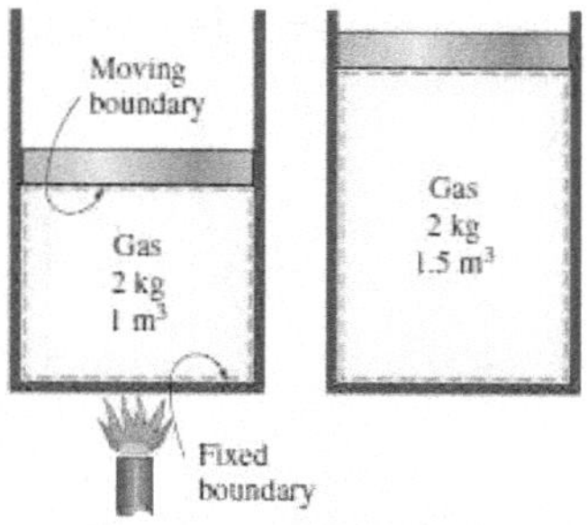

A closed system with a moving boundary.

An **open system**, or a **control volume**, as it is often called, is a properly selected region in space. It usually encloses a device that involves mass flow such as a compressor, turbine, or nozzle. Flow through these devices is best studied by selecting the region within the device as the control volume. Both mass and energy can crosstheboundaryofa control volume. The boundariesof a control volume are called a *control surface,* and they can be real or imaginary.

A large number of engineering problems involve mass flow in and out of a system and, therefore, are modelled as *control volumes.*

Example: A water heater, a car radiator, a turbine, and a compressor all involve mass flow and should be analyzed as control volumes (open systems) instead of as control masses (closed systems).

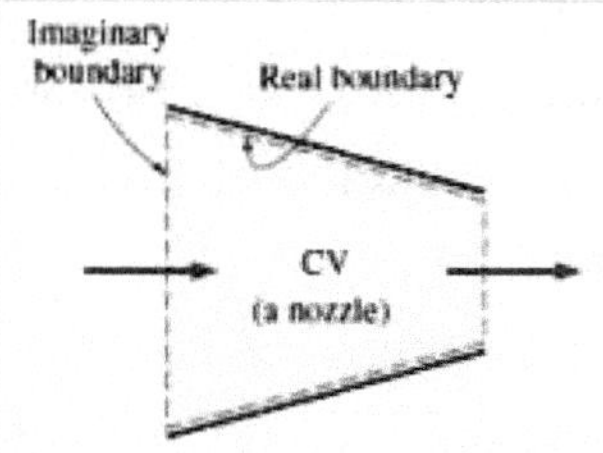

A control volume (CV) with real and imaginary boundaries

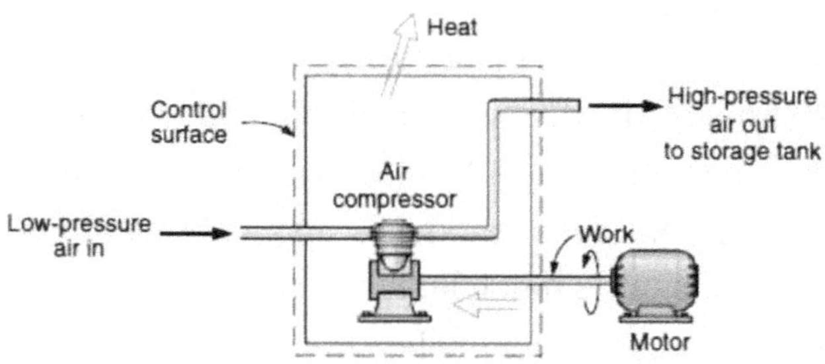

- **Properties:** The state or physical condition of the system can be described bysome parameters called as Properties. Properties are the coordinates to describe the state of the system. Every property has a fixed value at a particular state.
  Some familiar properties are pressure *P*, temperature *T*, volume *V*, and mass *m*. The list can be extended to include less familiar ones such as viscosity, thermal conductivity, modulus of elasticity, thermal expansion coefficient, electric resistivity, and even velocity and elevation.
- Propertiesareconsideredtobeeither*intensive*or *extensive.*
- **Intensive properties are those** that are independent of the Size or Extent or mass of a system,

such as temperature, pressure, and density.

- **Extensive properties** are those whose values depend on the size—or extent—of the system. Total mass, total volume, and total momentum are some examples of extensive properties. An easy way to determine whether a property is intensive or extensive is to divide the system into twoequalpartswithanimaginarypartition,as shown in Fig.Eachpartwillhavethesame valueofintensivepropertiesastheoriginalsystem,buthalfthe valueoftheextensive properties.

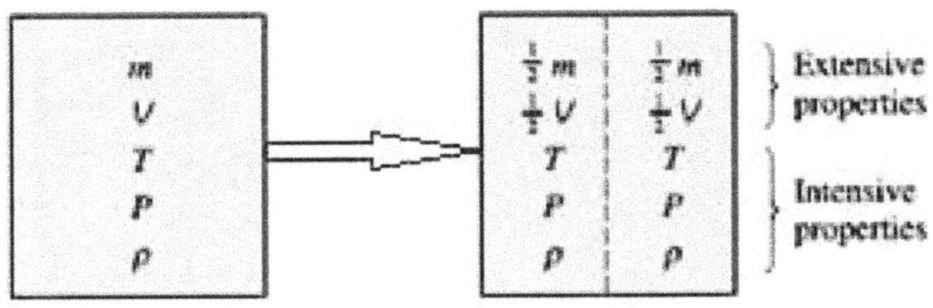

Criterion to differentiate intensive and extensive properties.

- Extensive properties per unit mass are called **specific properties**. Some examples of specific properties are specific volume ($v = V/m$) and specific total energy ($e = E/m$).
- **Continuum:** Matter is made up of atoms that are widelyspaced in the gas phase. Yet it is very convenient to disregard the atomic nature of a substance and view it as a continuous, homogeneous matter with no holes, that is, a **continuum**.
- The continuum idealization allows us to treat properties as point functions and to assume the properties vary continually in space with no jump discontinuities. This idealization is valid as longas the size of thesystem wedeal with is large relative to the space between the molecules.

- To have a sense of the distance involved at the molecular level, consider a container filled with oxygen at atmospheric conditions. The diameter of the oxygen molecule is about
- 3×10^{-10}m and its mass is 5.3×10^{-26} kg. Also, the *mean free path* of oxygen at 1 atm pressure and 20°C is 6.3×10^{28} m. That is, an oxygen molecule travels, on average, a distance of 6.3×10^{28} m (about 200 times of its diameter) before it collides with another molecule.
- Also, there are about 3×10^{16} molecules of oxygen in the tiny volume of 1 mm^3 at 1 atm pressure and 20°C. The continuum model is applicable as long as the characteristic length ofthe system (such as its diameter) is much larger than the mean free path of the molecules.
- Atveryhighvacuumsorveryhighelevations,themeanfreepathmaybecomelarge(for example,itisabout0.1 mforatmosphericairatanelevationof100km).Forsuchcasesthe **rarefiedgasflowtheory**shouldbeused,andtheimpactofindividualmoleculesshouldbe considered. In this text we will limit our consideration to substances that can be modelled as a continuum.

➢ **State:**Thestateofasystemistheconditionofthesystemdescribedbythe valuesofitsproperties.

➢ **Equilibrium:**IfthePropertiesareinvariantwithrespecttotimethenthatstateiscalledasan Equilibrium state.

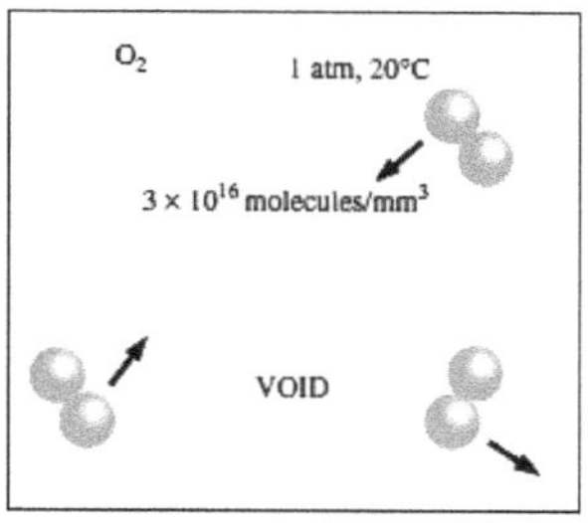

Despite the relatively large gaps between molecules, a gas can usually be treated as a continuum because of the very large number of molecules even in an extremely small volume.

Consider a system not undergoing any change. At this point, all the properties can be measured or calculated throughout the entire system, which gives us a set of properties that completely describesthe condition, or the **state**, of the system. At a given state, all the properties of a system have fixed values. If the value of even one propertychanges, the state will change to a different one. In the below figure shows a system at two different states.

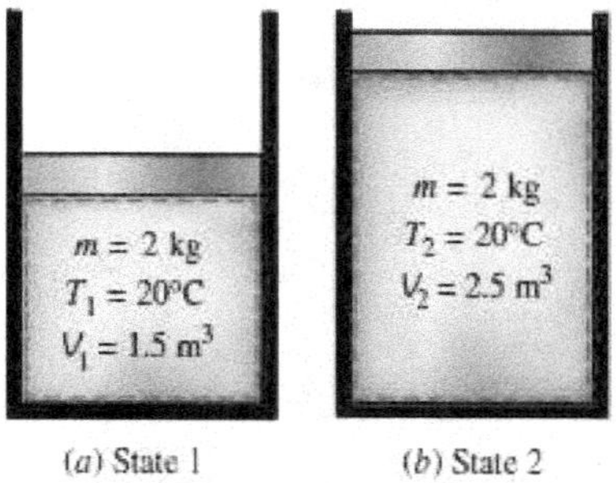

A system at two different states.

There are many types of equilibrium, and a system is not in **thermodynamicequilibrium** unless the conditions of all the relevant types of equilibrium are satisfied. For example,

iii. asystem is in **thermal equilibrium** ifthetemperatureis thesamethroughout theentiresystem, as shown in Fig. That is, the system involves no temperature differential, which is the driving force for heat flow.

iv. **Mechanical equilibrium** is related to pressure, and a system is in mechanical equilibrium if there is no change in pressure at any point of the system with time. However, the pressure may vary within the system with elevation as a result of gravitational effects.

v. Finally, a system is in **chemical equilibrium** if its chemical composition does not change with time, that is, no chemical reactions occur. A system will not be in equilibrium unless all the relevant equilibrium criteria are satisfied.

If all the above equilibrium are satisfied then the system is said to be in Thermodynamic Equilibrium.

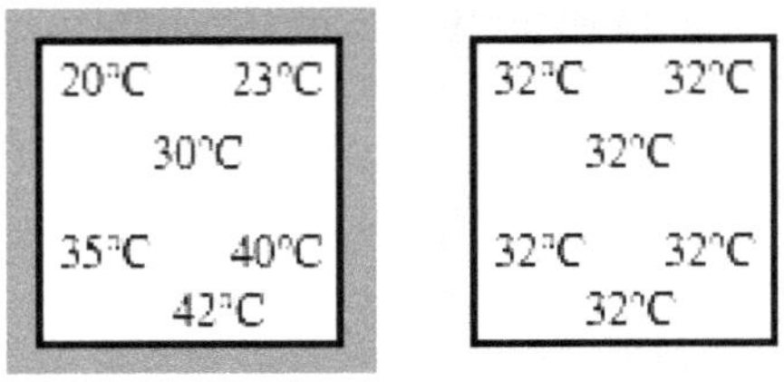

(a) Before (b) After

A closed system reaching thermal equilibrium.

StatePostulate:

- The state of a simple compressible system is completely specified by two independent, intensive properties.

Processesandcycles:

- Anychangethatasystemundergoesfromoneequilibriumstatetoanotheriscalleda **process**,andtheseriesofstatesthroughwhichasystempassesduringaprocessis called the **path** of the process.

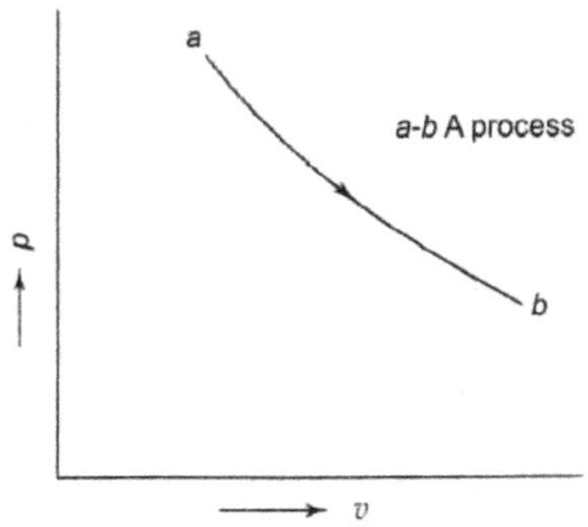

- Todescribeaprocesscompletely,oneshouldspecifytheinitialandfinalstatesofthe process, as well as the path it follows, and the interactions with the surroundings.
  Iftheprocessproceedsin suchawaythatif thefinalstateissameasinitialstatethen a

CYCLEissaidtobecompleted.

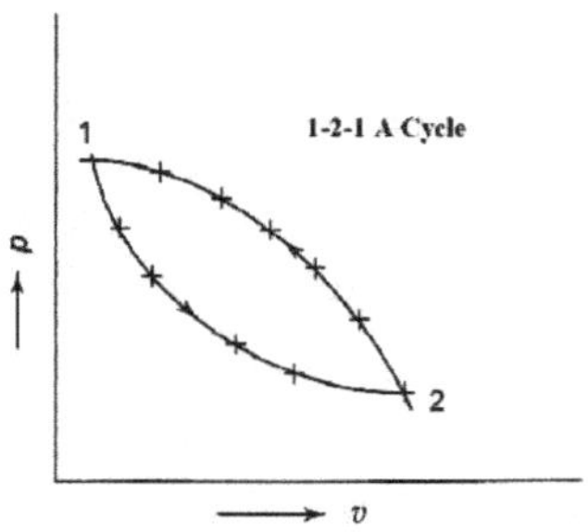

- **QuasistaticProcess**
  Whenaprocessproceeds insuchamannerthatthesystemremains infinitesimally closetoanequilibriumstateatalltimes,itiscalleda**quasi-static**,or**quasi-equilibrium**, **process**.

- **ExpansionorCompressionWorkinQuasi-EquilibriumProcesses**

AQuasiEquilibriumprocessisoneinwhichthedeparturefromthermodynamic equilibriumis atmostinfinitesimal.Allstatesthrough whichthe systempassesin aquasiequilibriumprocess maybeconsideredequilibriumstates. Becausenonequilibriumeffectsareinevitably present duringactualprocesses,systemsofengineeringinterestcanatbestapproach,butneverrealize,a quasi equilibriumprocess.

Imaginethatoneofthemassesisremoved,allowingthepistontomoveupwardasthegasexpands slightly. During such an expansion, the state of the gas would depart only slightly from equilibrium. Thesystem wouldeventuallycome to anew equilibrium state, wherethe pressure and allother intensive properties would again be uniform in value. Moreover, were the mass replaced, the gaswouldberestoredtoitsinitialstate,while again the departure from equilibrium would be slight. If several of the masses were removed one after another, the gas would pass through a sequence of equilibriumstateswithouteverbeingfarfromequilibrium.Inthelimitastheincrementsofmass aremadevanishingly small,thegaswouldundergo a quasiequilibriumexpansionprocess.Aquasi equilibrium compression can be visualized with similar considerations.

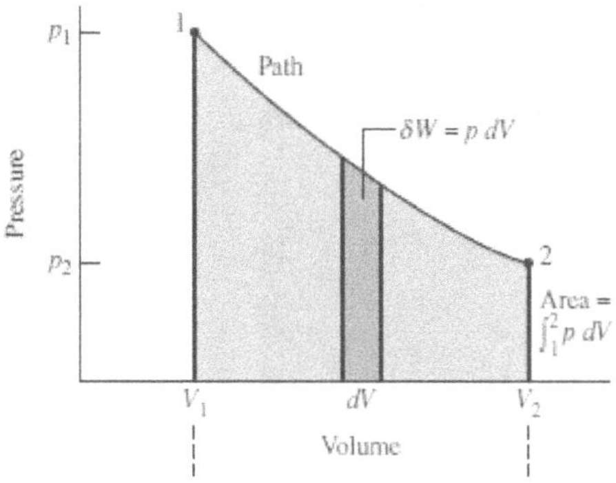

Work of a quasiequilibrium expansion or compression process.

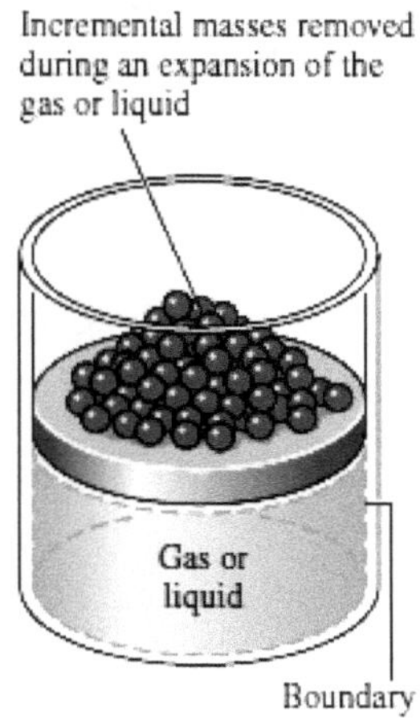

Illustration of a quasiequilibrium expansion or compression.

TEMPERATURE AND THE ZEROTH LAW OF THERMODYNAMICS

Temperature is rooted in the notion of the "hotness" or "coldness" of objects. We use our sense of touch to distinguish hot objects from cold objects and to arrange objects in their order of "hotness," deciding that 1 is hotter than 2, 2 hotter than 3, and so on. But however sensitive human touch maybe, we are unable to gauge this quality precisely.

To illustrate this, consider two copper blocks, and suppose that our senses tell us that one is warmer thantheother. If the blockswerebrought intocontactandisolatedfromtheirsurroundings,theywould interact in a way that can be described as a thermal (heat) interaction. During this interaction, it would be observed that the volume of the warmer block decreases somewhat with time, while the volume of the colder block increases with time. Eventually, no further changes in volume would be observed, and the blocks would feel equally warm. Similarly, we would be able to observe that the electrical resistance of the warmer block decreases with time and that of the colder block increaseswith time; eventually the electrical resistances would become constant also. When all changes in such observable properties cease, the interaction is at an end. The two blocks are then in thermal equilibrium. Considerations such as these lead us to infer that the blocks have a physical propertythat determines whether they will be in thermal equilibrium. This property is called **temperature**,andwepostulatethatwhenthetwoblocksareinthermalequilibrium, their temperatures are equal.

Zeroth Law of Thermodynamics:

- when two objects are in thermal equilibrium with a third object, they are in thermal equilibriumwith one another.
- If we want to know if two objects are at the same temperature, it is not necessaryto bringthem into contact and see whether their observable properties change with time, as described previously. It is necessary only to see if they are individually in thermal equilibrium with a third object. The third object is usually a *thermometer*.
- Any object with at least one measurable property that changes as its temperature changes can be used as a thermometer. Such a property is called a **thermometric property**. The particular substancethatexhibitschangesinthethermometricpropertyisknownas a *thermometricsubstance.*

Temperature scales:

- The temperature scales used in the SIand in the English system today are the **Celsius scale** and the **Fahrenheit scale** respectively**.** On the Celsius scale, the ice and steam points were originally assigned the values of 0 and 100°C, respectively. The corresponding values on the Fahrenheit scale are 32 and 212°F. These are often referred to as *two-point scales* since temperature values are assigned at two different points.
- In thermodynamics, it is very desirable to have a temperature scale that is independent of the properties of any substance or substances. Such a temperature scale is called a **thermodynamic temperature scale**, which is developed later in conjunction with the second law ofthermodynamics.
- The thermodynamic temperature scale in the SI is the **Kelvin scale**, named after Lord Kelvin (1824–1907). Thetemperatureunit on this scaleis the **kelvin**, which is designated byK. Thelowest temperature on the Kelvin scale is absolute zero, or 0 K.
- The thermodynamic temperature scale in the English system is the **Rankine scale**, named after William Rankine (1820–1872). The temperature unit on this scale is the **rankine**, which is designated by R.
- The relationship of the Kelvin, Rankine, Celsius, and Fahrenheit scales is shown in Fig. together with values for temperature at three fixed points: the triple point, ice point, and steam point.
- By international agreement, temperature scales are defined by the numerical value assigned to the easily reproducible **triple point** of water: the state of equilibrium among steam, ice and liquidwater.Thetemperatureatthisstandardfixedpointisdefinedas273.16Kelvins,abbreviatedas 273.16 K.

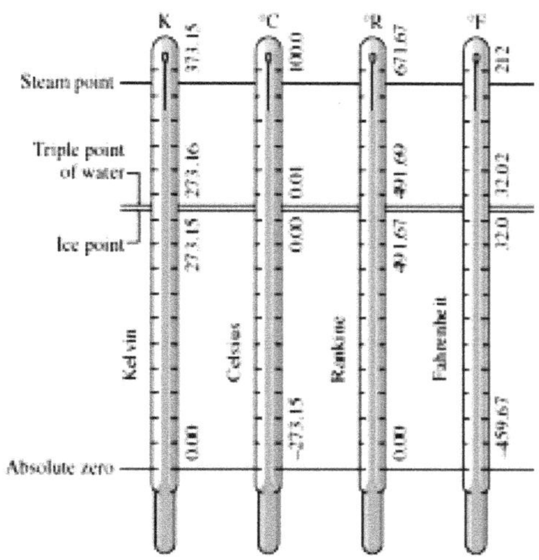

Comparison of temperature scales.

- Therelationshipbetween centigradescaleandKelvinscaleis$T(^0C)=T(K)-273.15$.
- Therelationshipbetween fahrenheitscaleandRankinescaleis$T(^0F)=T(R)-459.67$.
- Therelationshipbetween fahrenheitscaleandcentigradescaleis$T(^0F)=1.8T(^0C)+32$

Constant volume Gas thermometer:

- A small amount of gas is enclosed in bulb B which is in communication via the capillary tube C with one limb of the mercury manometer M. The other limb of mercury manometer is open to the atmosphere and can be moved vertically to adjust the mercury levels so that mercury just touches the Lip L of the capillary. The pressure in the bulb is used as a thermometric property and is given by

$p = p_o + \rho_m Z g$.

wherep$_o$istheatmosphericpressureand$\rho_m$isthedensityofmercury.

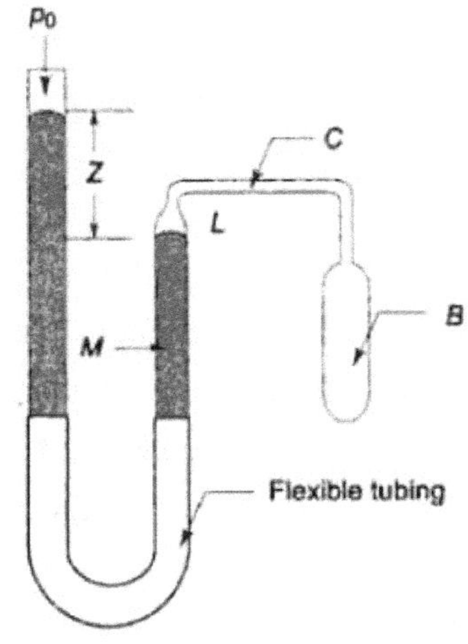

Constant volume gas thermometer

when the bulb is brought in contact with the system whose temperature is to be measured, the bulb, in course of time, comes in thermal equilibrium with the system. The gas in the bulb expands, on being heated, pushing the mercury downwards. The flexible limb of the manometer is then adjusted so that the mercury again touches the lip L. The difference in the mercury levels Z is recorded and thepressure p of the gas in the bulb is estimated. since the volume of the trapped gas remains constant, from the ideal gas equation

$$\Delta T = \frac{V}{R} \Delta P$$

i.etemperatureincreaseisproportionaltopressure increase.

PointandPathfunctions:

- Thermodynamic quantities can be divided into two categories: point or state functions and path or process functions.

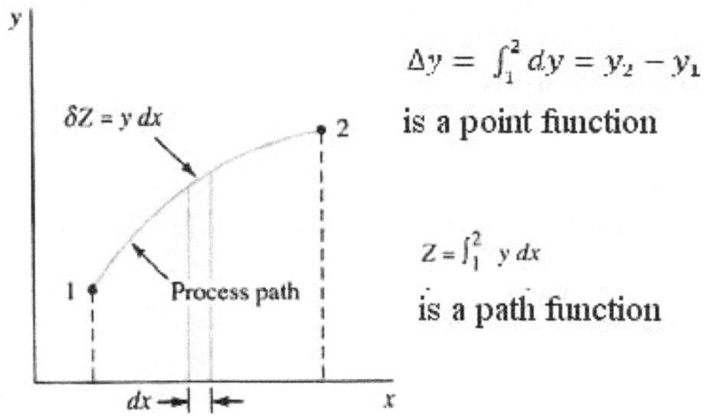

- All properties are state functions, since the change in the value of a property depends solely on the property values at the initial and final states, and not on the path of the process.
- A property change for a process with given end states 1 and 2 is described mathematically bythe exact differential *dy*, such that

$$\Delta y = \int_1^2 dy$$

- Quantitieswhosevalues dependonthepathofthe processarecalledprocessorpathfunctions.
- Pathfunctionsareinexactdifferentialsandhence

$$Z_{12} \neq \int_1^2 dz$$

Thetwomajorprocessfunctionsinthermodynamicstudiesareworkandheat.

Density,SpecificVolume,andSpecificGravity:

- Thedensityρisdefinedasthemassperunitvolume:

$\rho = \frac{mass}{volume} = \frac{m}{v}$,itsunitiskgperm^3

- Thespecificvolumevisdefinedasvolumeperunitmass:

$v = \frac{volume}{mass} = \frac{V}{m}$,itsunit ism^{3}perkg

- Thespecificgravityofasubstanceistheratioofitsdensitytothatofwaterataspecified temperature, such as 4 or 20°C and one atmosphere:

$$Sp.gr= \frac{\rho}{\rho_{water}}$$ ithasno units

- Thedensityofwaterat4°Ciscloseto1.00g/cm^3(kg/L),1000kg/m^3,whilethatofliquid mercury is close to 13.6 g/cm^3, 13600 kg/m^3.

- Thespecificweight*w*ofasubstanceistheweight perunit volume:

$$w=\frac{weight}{volume}=\frac{W}{V}$$,itsunitisNperm^3

Thermodynamicwork

- Work issaidtobe doneby a system if the sole effecton things externalto the system can bereduced to the raising of a weight.

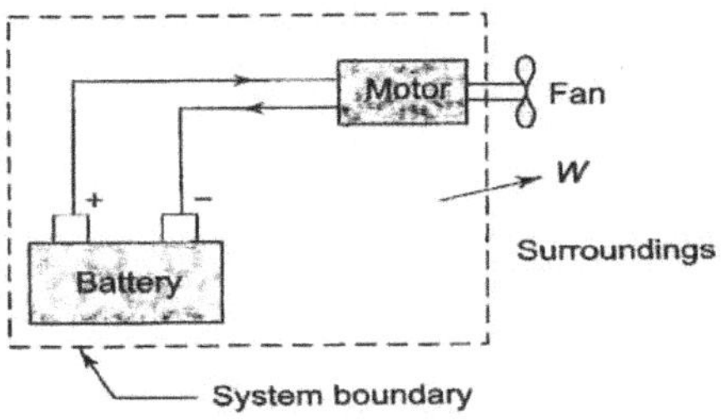

Battery-motor System driving a fan

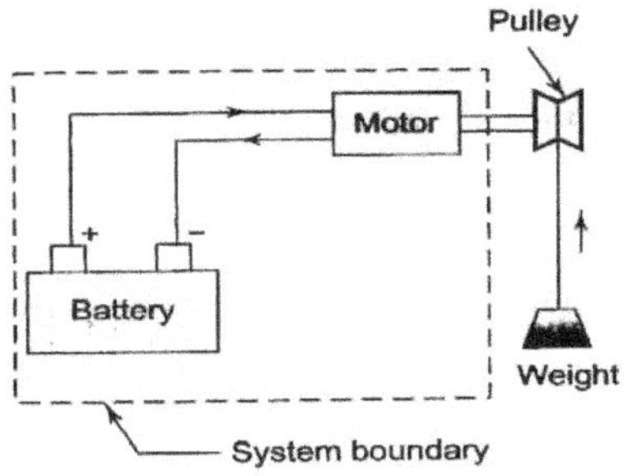

WorktransferfromaSystem Sign

Convention:

- When work is done bya system it is arbitrarily taken to be positive, and when work is done ona system, it is taken to be negative.

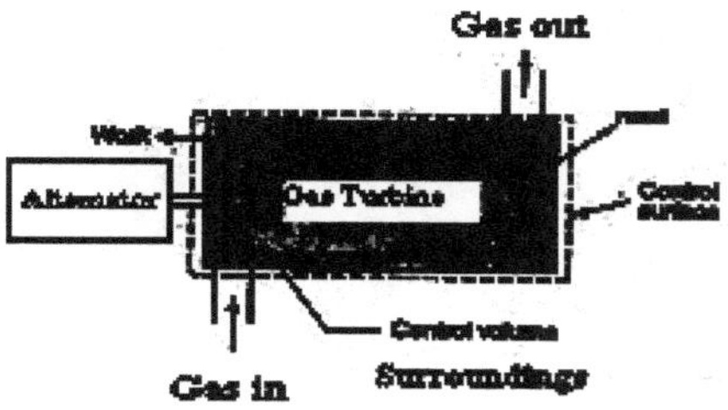

(a)Wis positive

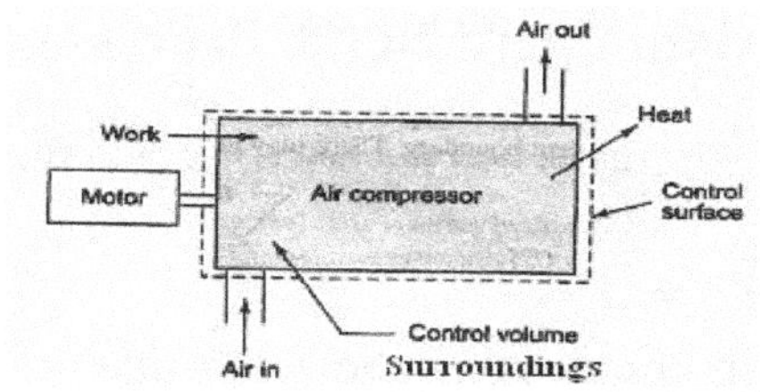

(b)W is negative

WorkinteractionbetweenaSystemandtheSurroundings

- ThesymbolWisusedforworktransfer.TheunitofworkisN.morJoule.
- Therateatwhichtheworkisdoneby,orupon,thesystemisknownaspower.Theunitof power is J/s or Watt.

Displacement work or pdV Work:

- Considerexpansionofgasinapiston-cylinderassemblyasshownin fig. 1.13.
- Theexpansionofthegasbyoutwardmovementofthepistoncanberepresentedonp-V diagram as given in Fig.1.14.

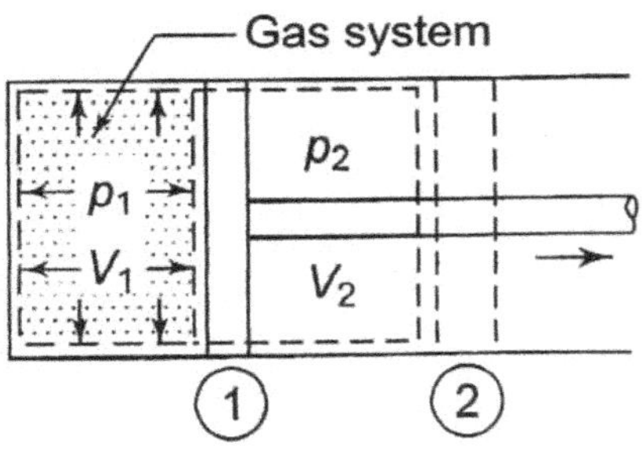

pdVwork

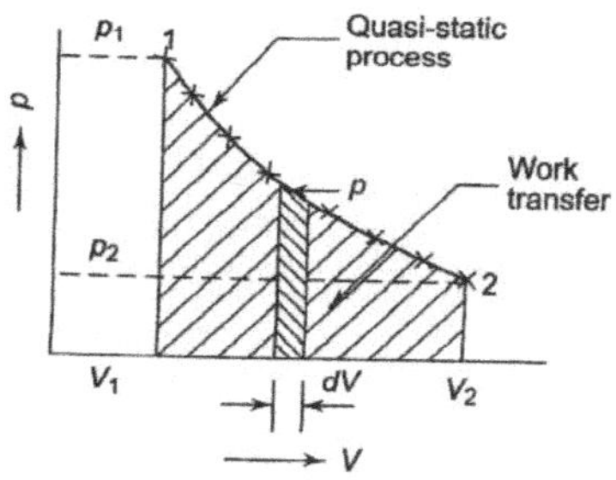

Quasi-staticpdVwork

- The shaded area represents work done due to small movement of the piston and the total work done by the gas during expansion process is

$$W_{1-2}=\int_1^2 pdV$$

- The integral $\int_1^2 pdV$ canbeevaluated,onlyifweknowtherelationbetween$p$andVorthe pathof the process.

- Theareaunderthecurve1-2representstheworkdoneduringtheprocess1-2.

pdV-work in Various Quasi-Static Processes:

- Constantpressureprocess(isobaricorisopiestic process)

$$W_{1-2}=\int_1^2 pdV=p(V_2-V_1)$$

- Constantvolumeprocess(isochoric process)

$$W_{1-2}=\int_1^2 pdV=0$$

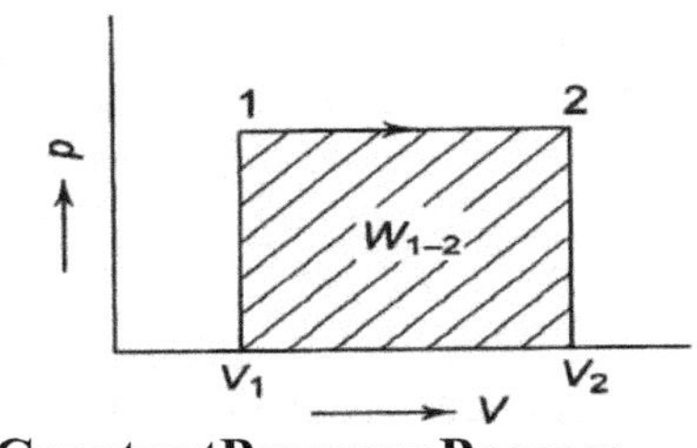

ConstantPressureProcess

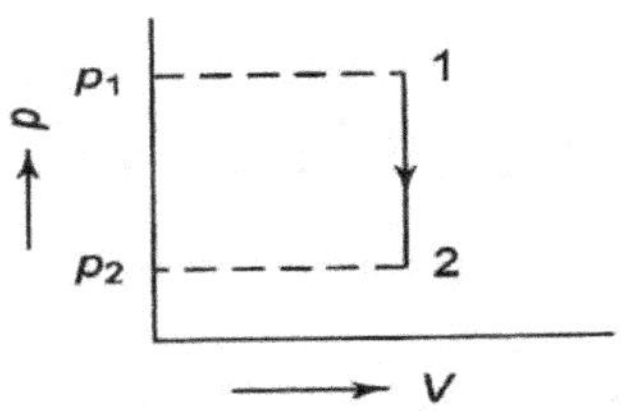

ConstantVolumeProcess

- ProcessinwhichpV=C

$$W_{1-2}=\int_{V1}^{V2} pdV= p_1v_1 ln\frac{V_2}{V_1}=p_1v_1 ln\frac{P_1}{P_2}$$

- ProcessinwhichpV^n=C,wherenisa constant

$$p_1v_1{}^n=p_2v_2{}^n$$

$$p=\frac{p_1v_1{}^n}{v^n}$$

$$W_{1-2}=\int_{V_1}^{V_2}\frac{C}{v^n}dv=C\left[\frac{V_2^{1-n}}{1-n}-\frac{V_1^{1-n}}{1-n}\right]=p_2v_2{}^n*\frac{V_2^{1-n}}{1-n}-p_1v_1{}^n*\frac{V_1^{1-n}}{1-n}$$

$$\therefore W_{1-2}=\frac{p_2v_2-p_1v_1}{n-1}$$

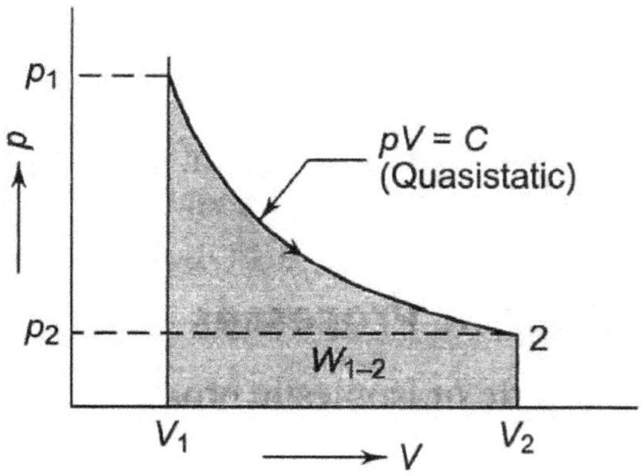

ProcessinwhichpV=Constant

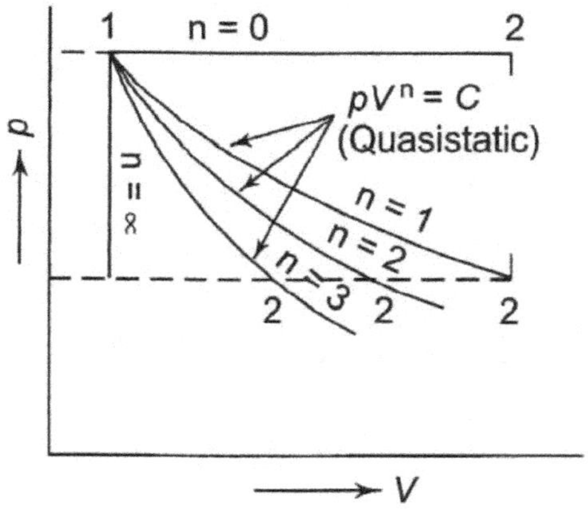

ProfessinwhichpV^n=Constant

Similarly,forprocessinwhichpv$^\gamma$=C

$$W_{1-2}=\frac{p_2v_2-p_1v_1}{\gamma-1}$$

OtherTypesofWorkTransfer

Electrical Work:

- Whenacurrentflowsthrougharesistor,takenasasystem,thereisworktransferintothe system.
- It is a thermodynamic work because the current can drive a motor, the motor can drive a pulley and the pulley can raise a weight.
- Thecurrentflow,*I*,inamperes,isgivenas

$I=\frac{dc}{dt}$ whereC is thechargeincoulombs and$t$ is timein seconds.

- If Eisthevoltagepotential,thework,$dW=E.dc=EIdt$
- $W=E.dc=\int_{t_1}^{t_2} EIdt$

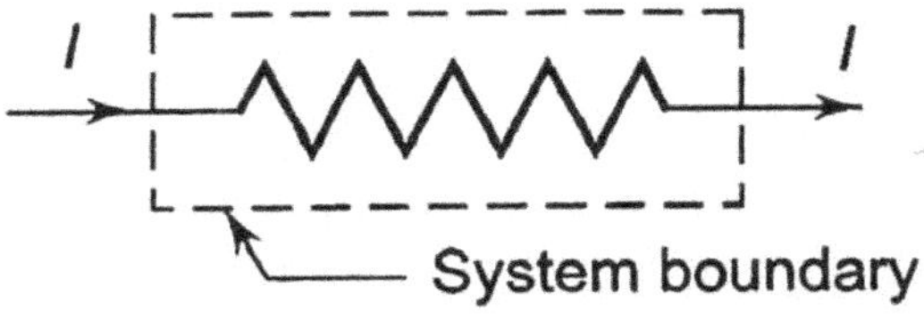

ElectricalWork

ShaftWork:

- Whenashaftrotatedbyamotor,takenasasystem,thereisworktransferintothesystem.

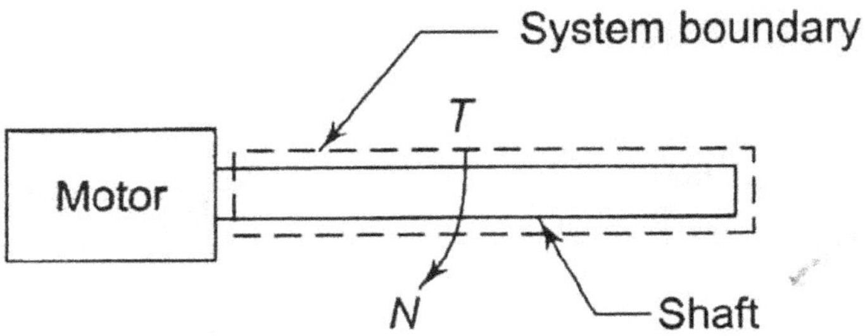

Shaftwork

- This is a form of thermodynamic work because the shaft can rotate a pulley which can raise a weight.
- If T is the torque applied to the shaft and$d\theta$is the angular displacement of the shaft, the shaft work is $W= \int_1^2 Td\theta$

SpringWork:

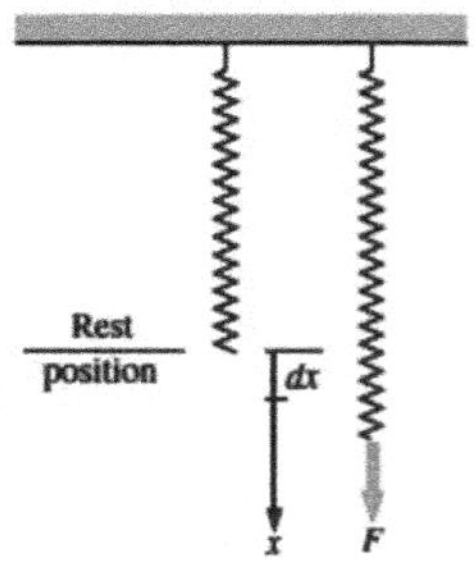

- Itiscommonknowledgethatwhenaforceisappliedona spring, the length of the spring changes.
- Whenthe lengthof the springchanges byadifferential amount dx undertheinfluenceof aforce$F$,theworkdoneis

$$dW_{spring}=Fdx$$

- Thedisplacementxisproportionaltotheforce applied

$$F=kx \qquad \text{(kN)}$$

- wherekisthespringconstantandhastheunitkN/m.Thedisplacement $x$is measuredfromthe undisturbed position of the spring (that is, $x = 0$ when F = 0).
- Work done=F.distance

ForInfinitesimalextension(dx)ofspringworkis$\delta w=F.dx$

$$\text{Total } work=\int_1^2 Fdx=\int_1^2 kxdx=k\left[\frac{x_2^2}{2}-\frac{x_1^2}{2}\right] \qquad \text{(kJ)}$$

1 1 ⌊2 2⌋

wherex₁andx₂aretheinitialandthefinaldisplacementsofthespring,respectively, measured from the undisturbed position of the spring.

Paddle-wheel work or Stirring Work:

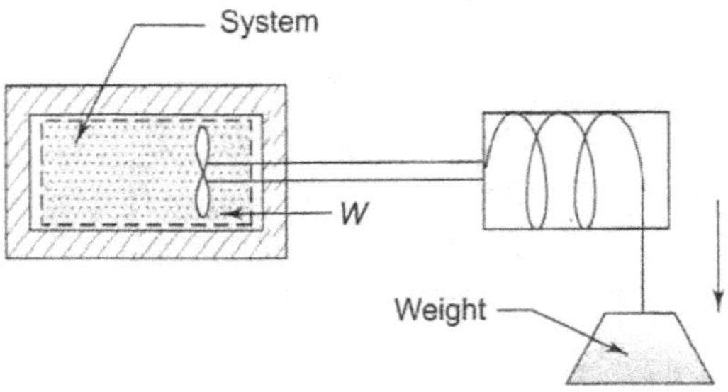

- Thepaddle-wheelturns whenthe weight is lowered.
- Thepaddle-wheelstirsfluidsystemand hence, thereisworktransferintothefluid system.
- Since the volume of the system remains constant, $\int_1^2 pdV = 0$
- If m is the mass of the weight lowered through a distance dz and T is the torque transmitted by the shaft in rotating through an angle d θ , the differential work transfer to the fluid is given by

dW=mgdz= $Td\theta = W = \int_1^2 mgdz = \int_1^2 Td\theta = 0$

- Thetotalworktransferis

$$mg\Delta z = T\Delta\theta$$

FlowWork:

- Theflowworkissignificantonlyinaflowprocessoranopensystem.
- When mass m enters or leaves a control volume, work is required to push the fluid into or outof the system. This term is referred to as flow work.
- Amass$m$inregion A(Fig.1.21)initiallyresidesjustoutsideacontrol surface.

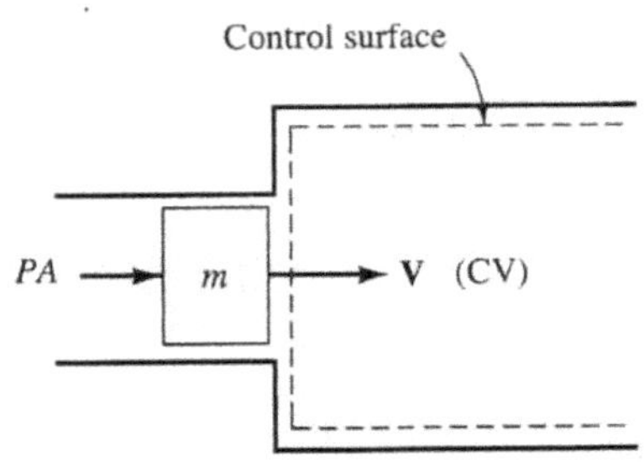

FlowworkataControlSurface

- The normal force exerted at the control surface of area A is simply the pressure times the area, or PA.
- Thustherateofflowwork$\dot{W}_{flow}$isgivenby

$\dot{W}_{flow}=pAV$,whereVisthevelocityoffluidmass$m$.

Since$m$=AV/v,therefore$\dot{W}_{flow}=pmv$.Where$v$isthespecificvolume.

- Theflowworkperunitmassispv.

Work done in Stretching a Wire:

- Letusconsiderawireasthe system.
- If the length of the wire is changed from L to L + dL, the infinitestimal amount of work that is done is equal todW = - F dL, where F is the tension in the wire.
- Theminussignisusedbecauseapositivevalueofd Lmeansanexpansionofthewire,for which work must be done on the wire, i.e., negative work.
- Forafinitechangeoflength.

$$W=-\int_1^2 Fdl$$

- Ifwelimittheproblemtowithintheelasticlimit,whereEisthemodulusofelasticity,τisthe stress,εisthestrain,andAisthecross-sectionalarea,then

$$F=\tau A=E\epsilon A, since E=\frac{\tau}{\varepsilon}$$

and$=\frac{dl}{L}$

$$dw=-FdL=-E\epsilon ALd\varepsilon$$

$$\therefore W=-A\varepsilon L\int^2 \epsilon d\epsilon=-\frac{AEL}{2(\epsilon^2-\varepsilon^2)}$$

Unrestrained or Free Expansion:

- The expansion of gas against a vacuum is called free expansion; essentially it is expansion which is not restrained by an opposing force.
- Let us consider a gas separated from the vacuum by a partition.Let the partition be removed. The gas rushes to fill the entire volume.

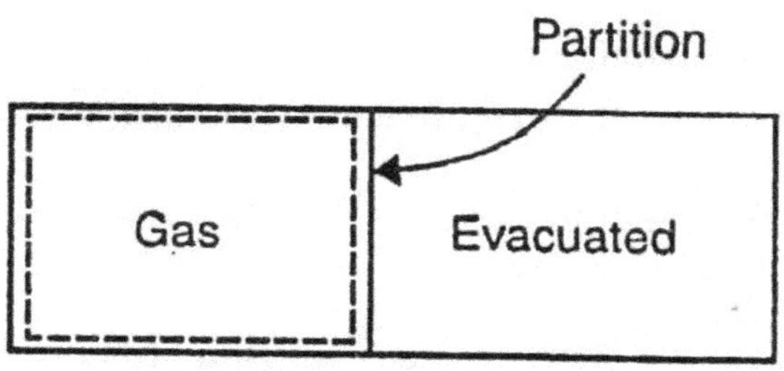

UnrestrainedExpansion

- Though this free expansion process involves a change of volume yet the work transfer is zero since the vacuum does not offer any resistance to the expansion process.
- Free expansion process is a totally irreversible process for which work transfer is zero even though $\int$ pdv is not zero.

Heat Transfer:

- Heat is defined as the form of energy that is transferred across a boundary by virtue of temperature difference.The temperature difference is the 'potential' or 'force' and heat transfer is the 'flux'.
- Heat flow into a system is taken to be positive, and heat flow out of a system is taken as negative.The symbol Q is used for heat transfer.
- Likeworktransfer,heattransferisapathfunctionandisaninexact differential.
- Aprocessinwhichnoheatcrossestheboundaryofthesystemiscalledanadiabaticprocess.
- Thus, an adiabatic process is one in which there is only work interactionbetween the systemand its surroundings.
- Awallwhichisimpermeabletotheflowofheatisanadiabaticwall,whereasawallwhich permits the flow of heat is a diathermic wall.

SpecificHeat:

- Thespecificheatofasubstanceisdefined astheamountofheatrequiredtoraiseaunitmassof the substance through a unit rise in temperature.The symbol *c* will be used for specific heat.
- For gases, iftheprocess is at constant pressure, it is c_p, and iftheprocess is at constant volume, it is c_v.
- Forsolidsandliquids,however,c_p=c_v=c,astheyareincompressible.

LatentHeat:

- Thelatent heat is theamount ofheat transfer required to causeaphasechange in unit mass ofa substance at a constant pressure and temperature.
- Therearethreephasesinwhichmattercanexist;solid,liquid,andvapourorgas.
- The latent heat of fusion (L_{fu}) is the amount of heat transferred to melt unit mass of solid intoliquid, or to freeze unit mass of liquid to solid.
- The latent heat of vaporization (L_{vap}) is the quantity of heat required to vaporise unit mass ofliquid into vapour or condense unit mass of vapour into liquid.
- The latent heat of sublimation (L_{sub}) is the amount of heat transferred to convert unit mass ofsolid to vapour or vice versa.
- L_{fu}isnotmuchaffectedbypressure,whereas L_{vap}ishighlysensitiveto pressure.

ComparisonofHeatandWork:

Similarities:

- Bothheatandworkarepathfunction;theyarenotpropertiesofthesystemandtheir differentials are not exact.
- Theyrepresentenergycrossingthesystemboundary;andhencetheyareaboundary phenomenon.
- Bothrepresenttransientphenomenon;theseenergyinteractionsoccuronlywhenasystem undergoes change of state.They exist as stored energy before or after the interaction.
- Conceptsofheatandworkareassociatednotwitha'store'butwitha'transferacross boundary'.

Dissimilarities:

- Therecannotbeworktransferinastablesystem. Nosuchrestrictionexistsforheat interaction.
- Heatisenergyinteractionduetotemperaturedifferenceonly;workisbyreasonsotherthan temperature difference.
- Heatisalowgradeenergywhereasworkisahighgradeenergy.

Unit – II

LearningMaterial

ThefirstLawofThermodynamicsforaControlMassUndergoingaCycle - Joule's Experiment

- Consider a closed system consisting of known mass of water, m, contained in an adiabatic vessel having a thermometer and a paddle wheel, as shown in fig.2.3.

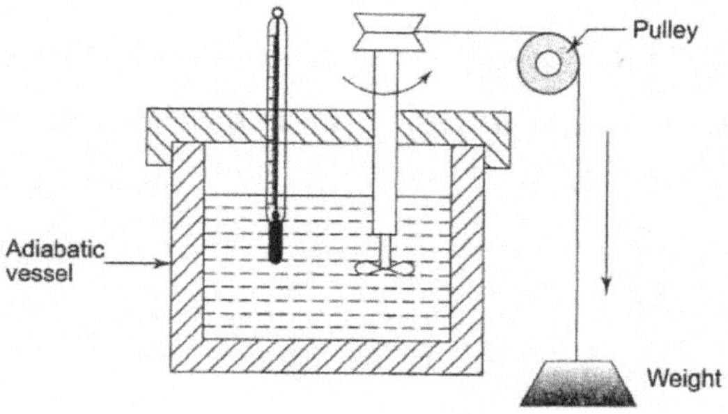

AdiabaticWork

- Let a certain amount of work W_{1-2}be done upon the system by the paddle wheel. The quantity of work can be measured by the fall of weight which drives the paddle wheel through a pulley.
- Let the system is initially at temperature T_1, the same as that of atmosphere, and after work transfer let the temperature rise to T_2.The pressure is always 1 atm.
- The process 1-2 undergone by the system is shown in fig. in generalized thermodynamic coordinates, X, Y.

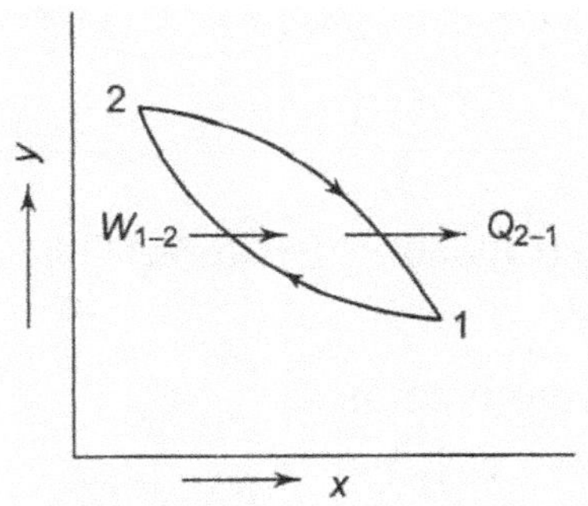

AdiabaticWorkTransferW_{1-2}FollowedbyHeatTransferQ_{2-1}

- Let the insulation now be removed.The system and the surroundings interact by heat transfer till the system returns to the original temperature T_1, attaining the condition of thermal equilibrium with the atmosphere.
- The amount of heat transfer $Q_{2-1}$from the system during this process, 2-1, shown in fig. can be estimated from $Q_{2-1} = m\, c_p(T_2 - T_1)$.
- The system thus executes a cycle, which consists of definite amount of work input

$W_{1-2}$to the system followed by the transfer of an amount of heat $Q_{2-1}$from the system. been found that this W_1-2is always proportional to the heat Q_2-1, and the constant of proportionality is called the Joule's equivalent or the mechanical equivalent of heat.

- In the simple example given here, there are only to energy transfer quantities asthe system performs a thermodynamic cycle.If the cycle involves many moreheat and work quantities, the same result will be found.Expressed algebraically.

$$(\Sigma W)_{cycle} = (\Sigma Q)_{cycle}$$

- Thisisalsoexpressedinthe form

$$\oint dW = J\oint dQ$$

- Where the symbol $\oint$denotes the cyclic integral for the closed path.This is the first law for a closed system undergoing a cycle.It is accepted as a general law of nature.Since no violation of it has ever been demonstrated.
- *The first law of thermodynamics states that during any cycle a system (Control mass) undergoes, the cyclic integral of the heat is equal to the cyclic integral ofthe work.*

The First Law of Thermodynamics for a Change in State of a Control Mass

- Consider a system that undergoes a cycle in which it changes from state 1 to state2 by process A and returns from state 2 to state 1 by process B.This cycle is shown in Fig. on a pressure – volume diagram.

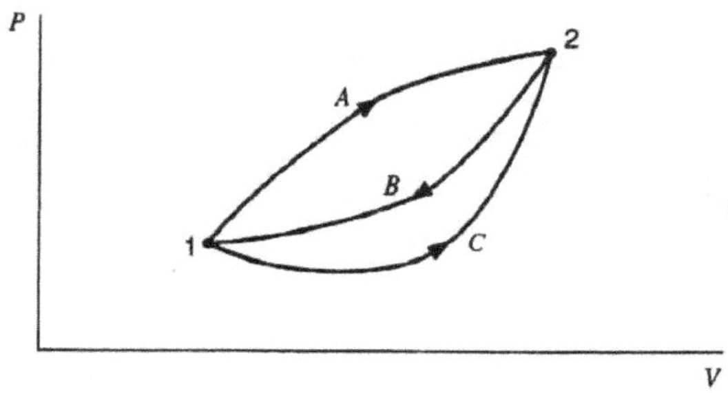

DemonstrationoftheExistenceofThermodynamicPropertyE

- Fromthefirstlawofthermodynamics

$$\oint \delta Q = \oint \delta W$$

- Considering the two separate processes, we have $\int_1^2 \delta Q_A + \int_2^1 \delta Q_B = \int_1^2 \delta W_A + \int_2^1 \delta W_B$
- Nowconsideranothercycleinwhichthecontrolmasschangesfromstate1to state 2by process Cand returns tostate1by processB, as before.For this cycle wecanwrite $\int_1^2 \delta Q_C + \int_2^1 \delta Q_B = \int_1^2 \delta W_C + \int_2^1 \delta W_B$

Subtracting the second of these equations from the first, we obtain $\int_1^2 \delta Q_A - \int_1^2 \delta Q_C = \int_1^2 \delta W_A - \int_1^2 \delta W_C$

or by rearranging,

$$\int_1^2 (\delta Q - \delta W)_A = \int_1^2 (\delta Q - \delta W)_C$$

- Since A and C represent arbitrary processes between states 1 and 2, the quantity $(\delta Q - \delta W)$ is the same for all processes between states 1 and 2. Therefore, $(\delta Q - \delta W)$ depends only on the initial and final states and not on the path followed between the two states. We conclude that this is a point function, and therefore it is the differential of a property of the mass. This property is the energy of the mass and is given the symbol *E*. Thus we can write dE $= \delta Q - \delta W$
- Because *E* is property, its derivative is written *dE*. When equation is integrated from an initial state 1 to a final state 2, we have $E_2 - E_1 = Q_{1-2} - W_{1-2}$
- Where E_1 and E_2 are the initial and final values of the energy E of the control mass, $Q_{1\text{-}2}$ is the heat transferred to the control mass during the process from state 1 to state 2, and $W_{1\text{-}2}$ is the work done by the control mass during the process.
- The physical significance of the property E is that it represents all the energy of the system in the given state. This energy might be present in a variety of forms, such as the kinetic or potential energy of the system as a whole with respect to the chosen coordinate frame, energy associated with the motion and position of the molecules, energy associated with the structure of the atom, chemical energy present in a storage battery, energy present in a charged condenser, or any of a number of other forms.
- E=Internal energy+kinetic energy+potential energy (or)
- E=U+KE +PE
- $dE = dU + d(KE) + d(PE) = \delta Q - \delta W$
- In words this equation states that as a control mass undergoes a change of state, energy may cross the boundary as either heat or work, and each may be positive or negative. The net change in the energy of the system will be exactly equal to the net energy that crosses the boundary of the system. The energy of the system may change in any or three ways – by a change in internal energy, in kinetic energy, or in potential energy.
- The kinetic energy is expressed as $KE = \frac{1}{2} m V^2$, where V is the velocity of the system as a whole.

- Thepotentialenergyisexpressedas$PE = mgZ$,whereZisthealtitudeofthe center of gravity of the system as a whole from the datum line.

- Therefore,thefistlawthermodynamicsforachangeofstateofthesystemis expressed as

$$U_2 - U_1 + \frac{m(V_2^2 - V_1^2)}{2} + mg(Z_2 - Z_1) = Q - W$$

- *The net change of the energy of the control mass during a process is always equal to the net transfer of energy across the boundary as heat and work.*

Internal Energy – A thermodynamic Property

- Internalenergyisanextensivepropertybecauseitdependsonthemassofthe system.Similarly, kinetic and potential energies are extensive properties.
- Thesymbol*U*designatestheinternalenergyofagivenmassofsubstance.
- Theinternalenergyperunitmass,u,istheintensivepropertyofthesystem.

The Thermodynamic Property Enthalpy

- Let us consider a control mass undergoing a quasi-equilibrium constant-pressure process, as shown in fig.

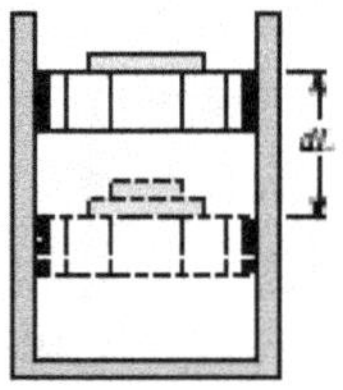

Constant-PressureQuasi-equilibriumProcess

- Assume that there are no changes in kinetic or potential energy and that the only work done during the process is that associated with the boundary (piston) movement.Taking the gas as control mass and applying the first law as

  $Q_{1-2} = U_2 - U_1 + W_{1-2}$

- Theworkdonecanbecalculatedfromtherelation$W_{1-2} = \int_1^2 PdV$
- Sincethepressureisconstant,$W_{1-2} = P\int_1^2 dV = P(V_2 - V_1)$
- Therefore,$Q_{1-2} = U_2 - U_1 + P_2V_2 - P_1V_1$

  $= (U_2 + P_2V_2) - (U_1 + P_1V_1)$

- We find that, in this very restricted case, the heat transfer during the process is given in terms of the change in the quantity (U + PV) between the initial and final states.

- Because all these quantities are thermodynamic properties, that is, functions only of the state of the system, their combination must also have these same characteristics.Therefore, we find it convenient to define a new extensive property, the enthalpy, H = U + PV or, per unit mass, h = u + pv

- The heat transfer in a constant-pressure quasi-equilibrium process is equal to the change in enthalpy, which includes both the change in internal energy and the work for this particular process.

The First Law as a Rate Equation

- $\underline{dU} + \underline{d(KE)} + \underline{d(PE)} = \dot{Q} - \dot{W}$

dt
dt

dt

- Intheabsenceofkineticandpotentialenergies, $\frac{dE}{dt} = \dot{Q} - \dot{W}$

Energyofanisolatedsystem

- Anisolatedsystemisoneinwhichthereisnointeractionofthesystemwiththe surroundings.
- ForanisolatedsystemdQ=0,dW=0.Hence,applicationfirst lawof thermodynamics gives dE = 0.Therefore, E = constant.
- Theenergyofanisolatedsystemisalwaysconstant.

PerpetualMotionMachineoftheFirstKind–PMM1

- There can be no machine which would continuously supply mechanical work without some other form of energy disappearing simultaneously.Such a fictitious machine is called a perpetual motion machine of the first kind, or in brief, PMM1. A PMM1 is thus impossible.

TheConstant-VolumeandConstant-PressureSpecificHeats

- The specific heat at constant volume is defined as the amount of heat required to raisethetemperatureof1kgsubstanceby1degreewhenvolumeismaintained constant $C_v = \frac{1}{m}\left(\frac{\delta Q}{\delta T}\right)_v$.
- Constantvolume,forwhichtheworkterm(PdV)iszero,sothatthespecificheat atconstantvolumecanalsobeexpressedas $C_v = \frac{1}{m}\left(\frac{\delta Q}{\delta T}\right)_v = \frac{1}{m}\left(\frac{\partial U}{\partial T}\right)_v = \left(\frac{\partial u}{\partial T}\right)_v$
- The specific heat at constant pressure is defined as the amount of heat required to raisethetemperatureof1kgsubstanceby1degreewhenpressureismaintained constant $C_p = \frac{1}{m}\left(\frac{\delta Q}{\delta T}\right)_p$
- Constant pressure, for which the work term can be integrated and the resulting PV termsattheinitialand finalstatescanbeassociatedwiththeinternalenergyterms, thereby leading to the conclusion that the heat transfer can be expressed in termsofthe enthalpy change.The correspondingspecific heat atconstantpressurecan beexpressedas $C_p = \frac{1}{m}\left(\frac{\delta Q}{\delta T}\right)_p = \frac{1}{m}\left(\frac{\partial H}{\partial T}\right)_p = \left(\frac{\partial h}{\partial T}\right)_p$

RelationshipbetweenSpecificHeatsandGasConstant

- Theenthalpy,h=u+ pv.

  = u+ RT

- dh=du+RdT

  Therefore, $\frac{dh}{dT}=\frac{du}{dT}\pm R$

  $$c_p=c_v+R;\qquad c_p-c_v=R$$

First-Law Analysis for a Control Volume

- Foranopensystem, massaswellasheatandworkcancrossthecontrolsurface, and the mass in the control volume can change with time.

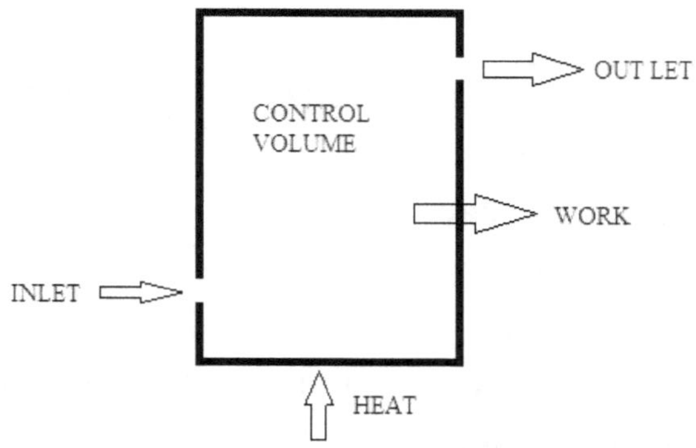

FlowProcessandControlVolume

- Applyingthelawofconservationofmasstothecontrolvolumewithseveral possible flows as

$$\frac{dm_{C.V.}}{dt}=\sum \dot{m}_i-\sum \dot{m}_e$$

- Theextensionofthefirstlawofthermodynamicsbecomes

$$\frac{dE_{C.V.}}{dt}=\dot{Q}_{C.V.}-\dot{W}_{C.V.}+\dot{m}_i(e_i+p_iv_i)-\dot{m}_e(e_e+p_ev_e)$$

$$=\dot{Q}_{C.V.}-\dot{W}_{C.V.}+\dot{m}_i\left(h_i+\frac{1}{2}V_i^2+gZ_i\right)-\dot{m}_e\left(h_e+\frac{1}{2}V_e^2+gZ_e\right)$$

- Foracontrolvolumewithseveralenteringandleavingmassflowrates,

$$\frac{dE_{C.V.}}{dt}=\dot{Q}_{C.V.}-\dot{W}_{C.V.}+\sum \dot{m}_i\left(h_i+\frac{1}{2}V_i^2+gZ_i\right)-\sum \dot{m}_e\left(h_e+\frac{1}{2}V_e^2+gZ_e\right)$$

Thesteady-StateProcess

- Thecontrolvolumedoesnotmoverelativetothecoordinateframe.
- Thestateofthemassateachpointinthecontrolvolumedoesnotvarywithtime.
- Asforthe massthatflowsacross thecontrolsurface,the massfluxandthestateof this mass at each discrete area of flow on the control surface do not vary withtime.

- Theratesatwhichheatandworkcrossthecontrolsurfaceremainconstant.
- Therefore, $\frac{dm_{C.V.}}{dt}=0$ andalso $\frac{dE_{C.V.}}{dt}=0$
- Thefirst-lawforacontrolvolumeundersteady-statewithseveralenteringand leaving mass flow rates is expressed as

$$Q_{C.V.} + \sum m_i \left(h_i + \frac{V_i^2}{2} + gZ_i\right) = \sum m_e \left(h_e + \frac{V_e^2}{2} + gZ_e\right) + W_{C.V.}$$

- Many of the engineering applications of the steady-state model involve only on flow stream entering and leaving the control volume.For this type of process, we can write

$$Q_{C.V.} + \dot{m}\left(h_i + \frac{V_i^2}{2} + gZ_i\right) = \dot{m}\left(h_e + \frac{V_e^2}{2} + gZ_e\right) + W_{C.V.}$$

Rearrangingthisequation,wehave

$$q + h_i + \frac{V_i^2}{2} + gZ_i = h_e + \frac{V_e^2}{2} + gZ_e + w$$

Where,$q$and$w$areheatandworkinteractionsperunitmassflowrate.

Examples of Steady-State Processes

a) HeatExchanger

- A heat exchanger is device in which heat transferred from one fluid to another. One such example is the steam condenser, where steam condenses outside the tubes and cooling water flows through the tubes, as shown in figure 2.5.

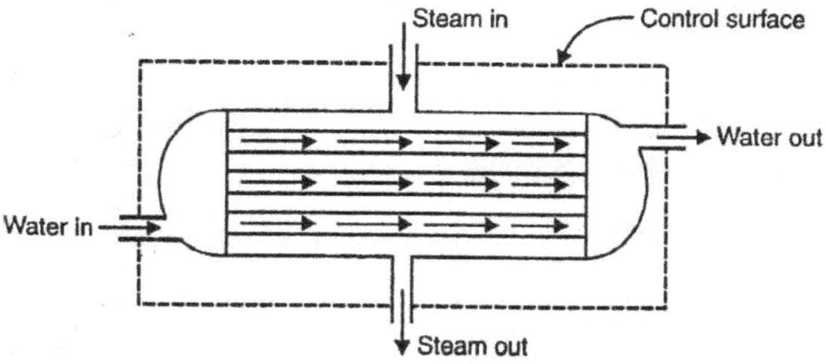

SchematicDiagramofaSteam Condenser

- Theflowthroughaheatexchangerischaracterizedby
  - (a) theHeatexchangeprocesstendstooccuratconstantpressureandfrictional pressure drop in the tubes is neglected.
  - (b) thereisnomeansfordoinganywork(shaftwork,electricalwork,etc.) and
  - (c) changesinkineticandpotentialenergiesarenegligiblysmall.
  - (d) noexternalheatinteraction.
- Using the subscripts '*s*' for steam and '*w*' for water and applying the steady flow energy equation for the heat exchanger, we get

$$\dot{m}_s(h_i)_s+\dot{m}_w(h_i)_w=\dot{m}_s(h_e)_s+\dot{m}_w(h_e)_w \quad \text{or}$$

$$\dot{m}_s(h_i-h_e)_s=\dot{m}_w(h_e- h_i)_w$$

b) Nozzleanddiffuser

- Anozzleisasteady-statedevicewhosepurposeistocreatehighvelocityfluid stream at the expense of the fluid pressure.
- Adiffuserisasteady-statedevicewhosepurposeistoraisethepressureofthe fluid at the expense of fluid velocity.

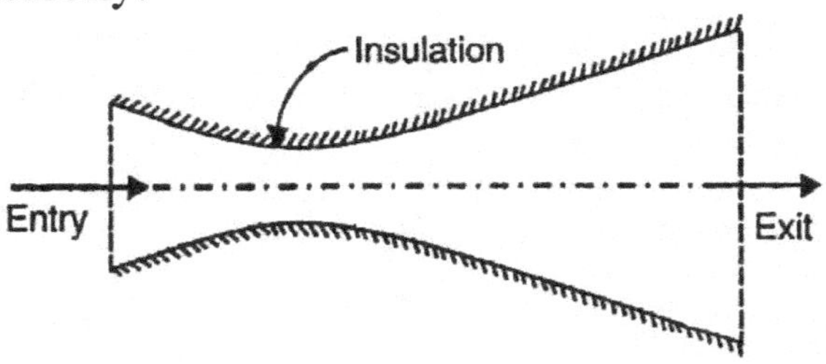

SchematicDiagramofaNozzle

- Theflowthroughanozzleischaracterizedbythefollowingfeatures:
  - (a) Thereisnomeanstodoanywork
  - (b) Thereislittleornochangeinpotentialenergy.
  - (c) Thereisusuallylittleornoheattransfer.An exception isthelarge nozzleon a liquid-propellant rocket.
- With these assumptions, the steady flow energy equation for the case of nozzle or diffuser becomes

$$h_i+\frac{V_i^2}{2}=h_e+\frac{V_e^2}{2}$$

c) SteamorGas Turbine

- A turbine is a rotary steady-state machine whose purpose is to produce shaft work at the expense of the pressure of the working fluid.Two general classes of turbines are steam turbines and gas turbines.

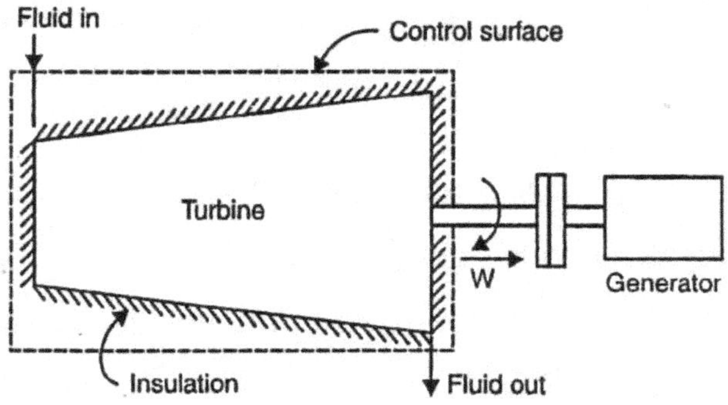

SchematicDiagramofa Turbine

- Thecharacteristicfeaturesofflowthroughasteamorgasturbineare:
  a) Thereislittleornochangeinpotentialenergy.
  b) Thereisusuallylittleornoheattransfer.
  c) Thereislittleornochangeinkineticenergy.
- Withtheseassumptions,thesteady flowenergyequationforthecaseofturbine becomes

$$\dot{m}h_i=\dot{m}h_e+\dot{W}_{shaft}$$

d) RotaryCompressor

- The purpose of the steadystate compressor is to increase the pressure of afluid by putting in shaft work.The most common is a rotary type compressor (either axial flow or radial / centrifugal flow).

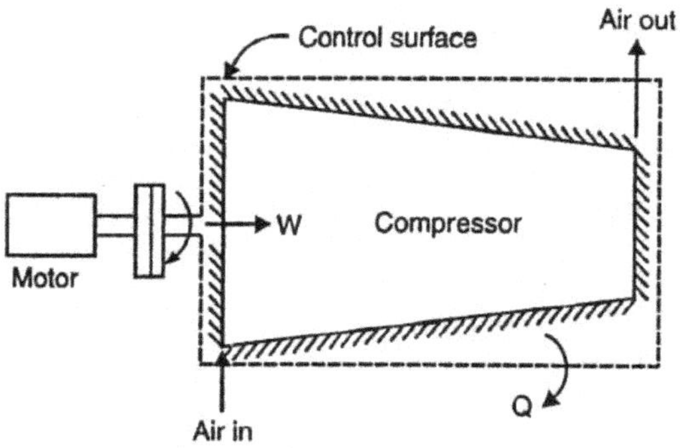

Schematic Diagram of a Rotary Compressor

- Thecharacteristicfeaturesofflowthrougharotarycompressorare
  a) Shaftworkisdoneonthesystemandthefluidiscompressed; Wisnegative.
  b) Heatislosttothesurroundings; Qisnegative
  c) Thereislittleornochangeinpotentialenergy.
- Withtheseassumptions,thesteadyflowenergyequationmaybewritten as

$$\dot{m}(h_i+\frac{V_i^2}{2})-\dot{Q}=\dot{m}(h_e+\frac{V_e^2}{2})-\dot{W}_{shaft}$$

If the velocity changes are neglected and the flow process is treated as adiabatic(Q=0), then

$$\dot{W}_{shaft}=\dot{m}(h_e-h_i)$$

e) ThrottlingDevice

- When a fluid flows through a restricted passage, like a partially opened wall, an orifice, or a porous plug, there is an appreciable drop in pressure, and the flow is said to be throttled.Figure 2.8 shows the process of throttling by a partially opened valve on a fluid following in an insulated pipe.

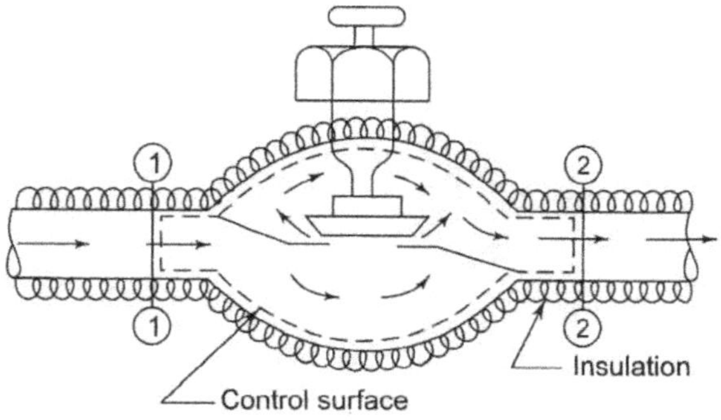

FlowthroughaValve

- Theflowthroughathrottlingdeviceischaracterizedbythefollowingfeatures:
  a) Thereisnomeanstodoanywork.
  b) Thereislittleornochangeinpotentialenergy.
  c) Thereisusuallylittleornoheattransfer.
- Withtheseassumptions,thesteadyflowenergyequationmaybewritten as

$$(h_i + \frac{V_i^2}{2}) = (h_e + \frac{V_e^2}{2})$$

- Often the velocities in throttling are so low that the kinetic energy terms are alsonegligible.So

$$h_i=h_e$$

- Therefore, throttling process is an isenthalpic process for which enthalpy remains constant.

Unit – III

LearningMaterial

LimitationsofFirstLawofThermodynamics

- Firstlawfixestheexchangeratebetween heat andwork, andplacesnorestrictions on the direction of change.
- Processes proceed spontaneously in certain directions, but the reverse is not automatically attainable even though the reversal of the processes does not violate the first law.
- First law provides a necessarybut not a sufficient condition for a process to occur, and
- There does exist some directional law which would tell whether a particular process occurs or not.Answer is provided by the second law of thermodynamics.

ThermalReservoir

- A thermal (heat) reservoir is that part of environment which can exchange heat energywith a system.It has sufficiently large heat capacity and its temperature is not affected by the quantity of heat transferred to or from it.
- A thermal reservoir is thus characterized by its temperature which remains constant.
- The reservoir which is at high temperature and supplies heat is known as heat source.Examples are a boiler furnace, a combustion chamber and a nuclear reactor, etc.
- Thereservoirwhich is atlowtemperatureand towhich heatistransferrediscalled the heat sink.Atmospheric air, ocean and river etc. constitute the heat sink.

HeatEngine

- A heat engine maybe defined as a device that operates in a thermodynamic cycle anddoesa certainamountof netpositiveworkthrough thetransferof heatfroma high-temperature body to a low-temperature body
- Theinternalcombustionengineandthegasturbineareexamplesofsuchdevices.
- Anotherexampleofaheatengineisthethermoelectricpowergenerationdevice.

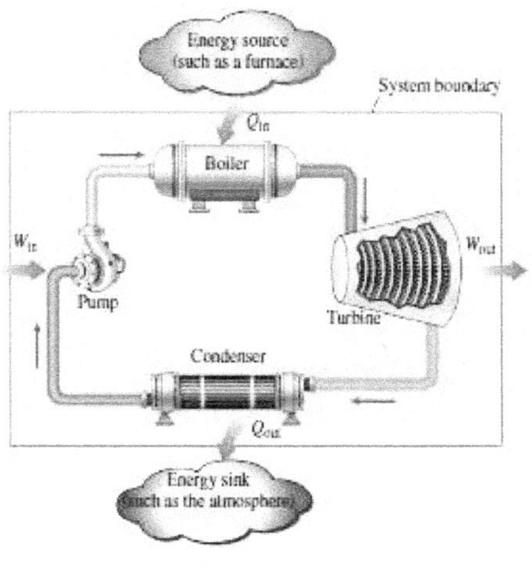

Fig.3.1:ASimpleSteamPowerCycle

- Thermalefficiencyisdefinedastheratioofnetworkoutputtoheatinput

$$\eta_{thermal} = \frac{Network output}{Heat\ input} = \frac{W_{net}}{Q_H} = \frac{Q_H - Q_L}{Q_H}$$

- Typical values for the thermal efficiency of real engines are about 35-50% for large power plant, 30-35% for gasoline engines, and 35-40% for diesel engines.

RefrigeratororHeatPump

- A refrigerator or heat pump may be define as a device that operates in a thermodynamiccycle andtransferheatfromalow-temperaturespace orbodytoa high-temperature space or body by receiving work energy as input.
- The purpose of the refrigerator is to maintain the space at a temperature lower than the surroundings by extracting heat from it.

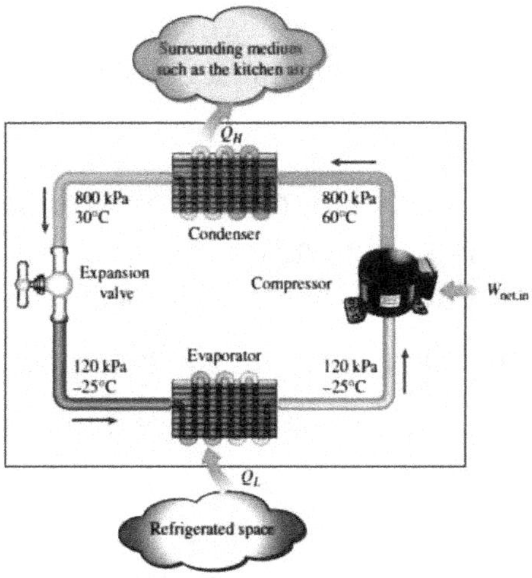

Fig.3.2:ASimpleRefrigerationCycle

- Thepurposeoftheheatpumpistomaintainthespaceatatemperaturehigher than the surroundings by supplying heat to it.
- The"efficiency"ofarefrigeratororheatpumpisexpressedintermsofthecoefficient of performance, which we designate as COP.
- Thecoefficientof performance isdefinedastheratioof desiredeffecttothework input.
- Forarefrigerator,thedesiredeffectistheamountofheatextractedfrom alow temperature space, Q_L. Thus, the COP of a refrigerator is

$$(COP)_R = \frac{Q_L}{W}$$

- Foraheatpump,thedesiredeffectistheamountofheatsuppliedtoahigh temperature space, Q_H. Thus, the COP of a heat pump is

$$(COP)_{HP} = \frac{Q_H}{W}$$

- TherelationshipbetweenCOPofheatpumpandCOPofrefrigeratorisderivedas

$$(COP) = \frac{Q_H}{W} = \frac{Q_L+W}{W} = 1 + \frac{Q_L}{W} = 1 + (COP)$$

HP W W W R

The Second Law of Thermodynamics

- **The Kelvin-Planck statement**:It is impossible to construct a device that will operate in a cycle and produce no effect other than the raising of a weight and the exchange of heat with a single reservoir.

- statement ties in with our discussion of the heat engine.It states that it is impossible to construct a heat engine that operates in a cycle, receives a given amountof heatfroma high-temperaturebody, anddoesanequalamountof work. The only alternative is that some heat must be transferred from the working fluid to a low-temperature body.
- Thisimpliesthatitisimpossibletobuilda heatengine that has thermal efficiency of 100%
- **The Clausius Statement**: It is impossible to construct a device that operates in a cycle and produces no effect other than the transfer of heat from a cooler body to a hotter body.
- This statement is related to the refrigerator or heat pump.It states that it is impossible to construct a refrigerator that operates without input of work.This also implies that the coefficient of performance is always less than infinity.

Perpetual-Motion Machine of Second Kind

- A perpetual-motion machine of the second kind would extract heat from a source and then convert this heat completely into other forms of energy, thus violating the second law.

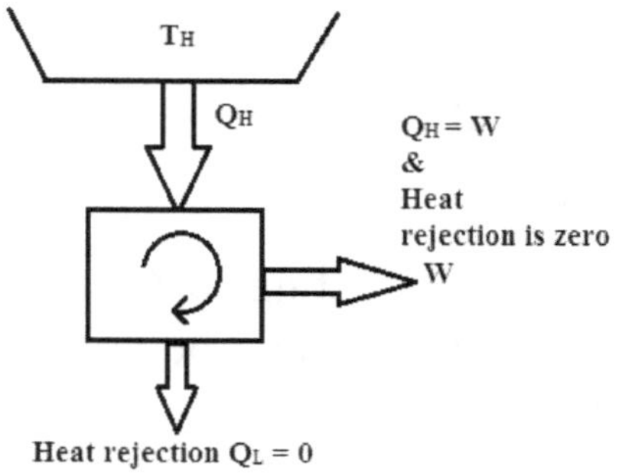

Fig.3.3:Perpetual-MotionMachineofSecondKind

- Perpetual-motionmachineofsecondkindhas100%thermalefficiency.

Equivalence of Kelvin Planck and Classius statement

AtfirstKelvin-Planck'sandclassiusstatementsmayappeartobeunconnected,butitcan easily be shown that they are virtually two parallel statements of the second law and are equivalent in all respects.

Theequivalenceofthetwostatementswillbeprovedifitcanbeshownthatthe violation of one statement implies the violation of the second, and vice versa.

1. Let us first consider a cyclic heat pump P which transfers heat from a low temperaturereservoir(T_2)toahightemperaturereservoir(T_1)withnoothereffect, i.e., with

no expenditure of work, ***violating classius statement***. Let us assume a cyclic heat engine E operating between the same thermal reservoirs, producing W_{net}in one cycle. The rate of working of the engine is such that it draws an amount of heat $Q_1$from the hot reservoir equal to that discharged by the heat pump. Then the hot reservoir maybe eliminated and the heat Q_1discharged bythe heat pump is fed to the heat engine. So we see that the heat pump P and the heat engineEactingtogetherconstituteaheatengineoperatingincyclesandproducing net work while exchanging heat only with one body at a single fixed temperature (T_2). This violates Kelvin-Planck Statement.

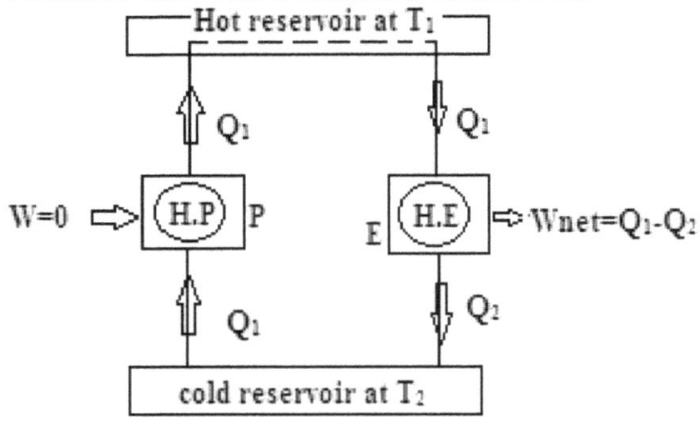

Violation of the classius statement

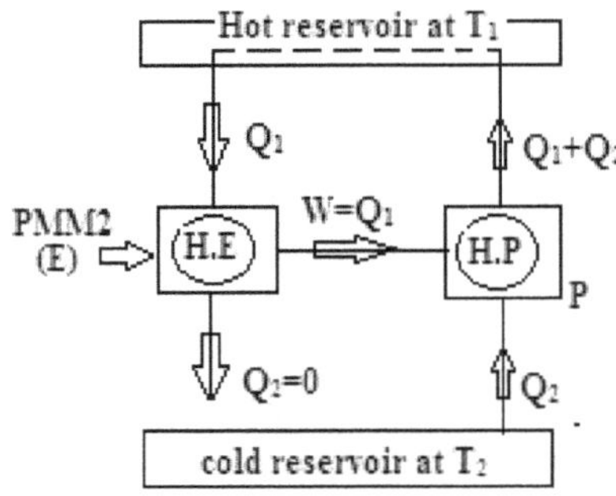

Violation of Kelvin-Planck statement

2. Let us now consider a perpetual motion machine of second kind (E) which produces net work in a cycle by exchanging heat with only one thermal reservoir (at T_1) and thus violates the Kelvin-planck statement. Let us assume a cyclic heat pump (P) extracting heat $Q_2$from a low temperature reservoir at $T_2$and dischargingheattoahightemperaturereservoiratT_1withtheexpenditureofwork W equal to what the PMM2 delivers in a complete cycle. So E and P together constitute a heat pump working in cycles and producing the sole effect of transferring heat from a lower to a higher temperature body, thus violating the Clausius statement.

Carnot Cycle and Carnot Heat Engine

A Carnot cycle is a hypothetical cycle consisting of four distinct processes: tworeversible isothermal processes and two reversible adiabatic processes.The cycle was proposed in 1824 by a young French engineer, Sadi Carnot.

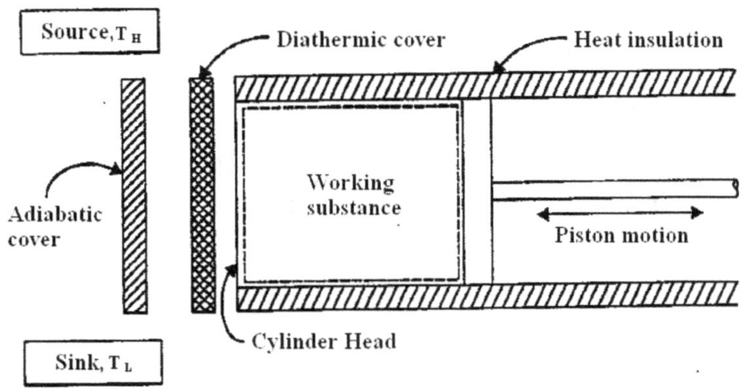

Fig. 3.4a:Essential Elements of a Heat Engine

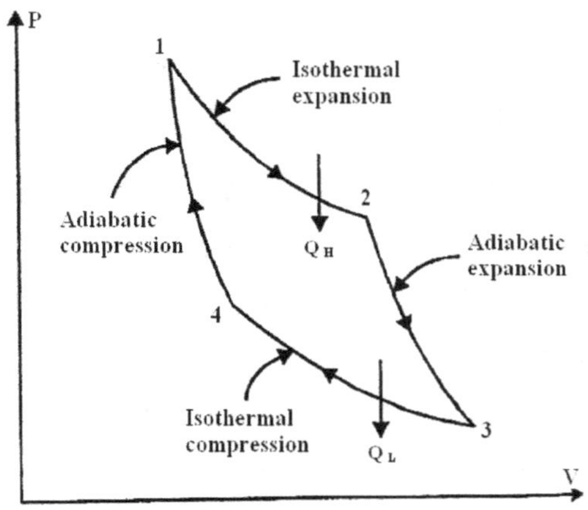

Fig.3.4b:CarnotHeatEngineCycle

working on Carnot Cycle **on P-V plot**

Thesequenceofoperationforthedifferentprocessesconstitutinga CarnotCycle is:

Isothermal expansion (1 – 2):The heat is supplied to the working fluid at constant temperature T_H.This is achieved by bringing the heat source in good thermal contact with the cylinder head through diathermic cover.The gas expands isothermally from state point 1 to state point 2.

The heat supplied equals the work done which is represented by area under the curve 1-2 on pressure-volume plot and is given by

$$Q_H = W_{1-2} = p_{a1} V \log_e \frac{V_2}{V_1} = m R T_1 \log_e \frac{V_2}{V_1}$$

Adiabatic Expansion (2-3):At the end of isothermal expansion (state point 2), the heat source is replaced by adiabatic cover.The expansion continues adiabatically and reversibly up to state point 3.Work is done by the working fluid at the expense of internal energy and its temperature falls to $T_L$at state point 3.

Isothermal Compression (3-4): After state point 3, the piston starts moving inwards and the working fluid is compressed isothermally at temperature T_L.The constant temperature $T_L$is maintained by removing the adiabatic cover and bringing the heat sink in contact with the cylinder head.The compression continues up to state point 4.

The working fluid loses heat to the sink and its amount equals the work done on the working fluid.This work is represented by area under the curve 3 - 4 and its amount is given by.

$$Q_L = W_{3-4} = p_3V_3 \log_e\frac{V_3}{V_4} = m\,R\,T_3 \log_e\frac{V_3}{V_4}$$

Adiabatic Compression (4 – 1): At the end of isothermal compression (state point 4),the heat sink is removed and is replaced by adiabatic cover.The compression now proceeds adiabatically and reversibly till the working fluid returns back to its initial state point 1.Work is done on the working fluid, the internal energy increases andtemperature is raised to T_H.

Since all the processes that constitute a Carnot cycle are reversible, the Carnot cycle is referredtoasareversiblecycle.ThethermalefficiencyofCarnotheatengineisgivenby

$$\eta = \frac{network output}{heat input} = \frac{W_{net}}{Q_H}$$

Therearenoheatinteractionsalongthereversibleadiabaticprocesses2–3and4–1, and application of the first law of thermodynamics for the complete cycle gives.

$$\oint \delta W = \oint \delta Q \quad \text{or}$$

$$W_{net} = Q_H - Q_L = mRT_1 \log_e\frac{V_2}{V_1} - mRT_3 \log_e\frac{V_3}{V_4}$$

Therefore, $$\eta = \frac{mRT_1\log_e\frac{V_2}{V_1} - mRT_3\log_e\frac{V_3}{V_4}}{mRT_1\log_e\frac{V_2}{V_1}} = 1 - \frac{T_3}{T_1} X \frac{\log_e\frac{V_3}{V_4}}{\log_e\frac{V_2}{V_1}}$$

Fortheadiabaticexpansionprocesses2–3and4–1,

$$\frac{T_2}{T_3} = \left(\frac{V_3}{V_2}\right)^{\gamma-1} and \frac{T_1}{T_4} = \left(\frac{V_4}{V_1}\right)^{\gamma-1}$$

SinceT_1=T_2=$T_H$andT_3=T_4=T_L,wehave $$\frac{V_3}{V_2} = \frac{V_4}{V_1} or \frac{V_3}{V_4} = \frac{V_2}{V_1}$$

Substitutingtheaboverelation3.10intheequation3.9,weget

$$\eta = 1 - \frac{T_3}{T_1} = 1 - \frac{T_L}{T_H} = \frac{T_H - T_L}{T_H}$$

FollowingconclusionscanbemadewithrespecttoefficiencyofaCarnotengine:

(1) Theefficiencyisindependentoftheworkingfluidanddependsuponthe temperatures of source and sink.

(2) Theefficiencyisdirectlyproportionaltotemperaturedifference(T_1–T_2) between the source and the sink.

(3) Higher the temperature differencebetweensourceandsink, thehigher will be the efficiency.

(4) Theefficiencyincreaseswithanincreaseintemperatureofsourceanda decrease in temperature of sink.

(5) IfT_1=T_2,noworkwillbedoneandefficiencywillbezero.

Metallurgical considerations and the high cost of temperature resisting materials limit the higher temperature T_1.The lower temperature $T_2$is limited by atmospheric or sink conditions.

Reversed Heat Engine (Carnot Heat Pump or Refrigerator)

Refrigeratorsandheatpumpsarereversedheatengines.

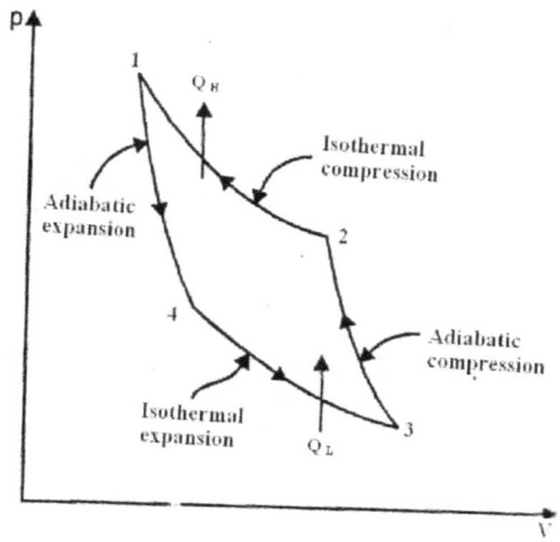

Fig.3.5:CarnotHeatPumporRefrigeratorCycleonP-Vplot

Fig. 3.5 shows the P –V plot ofa Carnot heat pump (i.e. reversedCarnot heat engine).The sequence of operation is:

1–4:Isentropic (reversible adiabatic) expansion of working fluid in the clearance space of the cylinder.The temperature falls from $T_1$and T_2.

4–3:Isothermal expansion during which heat $Q_L$is absorbed at temperature T_2fromthe space being cooled.

3–2:Isentropiccompressionofworkingfluid.ThetemperaturerisesfromT$_2$to T_1.

3–1:Isothermalcompressionofworkingfluidduringwhichheat$Q_H$isrejectedtoa system at higher temperature.

Asoutlinedabovei.e.inthecaseofheatengine,

$$Q_H = mRT_1 \log_e \frac{V_2}{V_1}; Q_L = mRT_3 \log_e \frac{V_3}{V_4}$$

Also $\frac{V_3}{V_4}=\frac{V_2}{V_1}$, $T_1=T_2=T_H$ and $T_3=T_4=T_L$

Therefore, for a heat pump

$$(COP)_{Heatpump}=\frac{Q_H}{Q_H-Q_L}$$

$$=\frac{mRT_1 log_e\frac{V_2}{V_1}}{mRT_1 log_e\frac{V_2}{V_1}-mRT_3 log_e\frac{V_3}{V_4}}=\frac{T_1}{T_1-T_3}=\frac{T_H}{T_H-T_L}$$

For a refrigerator,

$$(COP)_{Refrigerator}=\frac{Q_L}{Q_H-Q_L}$$

$$=\frac{mRT_3 log_e\frac{V_3}{V_4}}{mRT_1 log_e\frac{V_2}{V_1}-mRT_3 log_e\frac{V_3}{V_4}}=\frac{T_3}{T_1-T_3}=\frac{T_L}{T_H-T_L}$$

Carnot Theorem

No heat engine operating in a cycle between two given thermal reservoirs, with fixed temperatures, can be more efficient than a reversible engine (Carnot Engine) operating between the same thermal reservoirs.

ProofofCarnotTheorem

Consider a reversible engine $E_A$and an irreversible engine $E_B$operating between thesame thermal reservoirs at temperatures $T_H$and T_L.For the same quantity of heat Q_H withdrawn from the high temperature source, the work output from these engines is W_A and $W_B$respectively.As such the heat given off by the reversible engine is $(Q_H - W_A)$ and that from irreversible engine is $(Q_H - W_B)$.

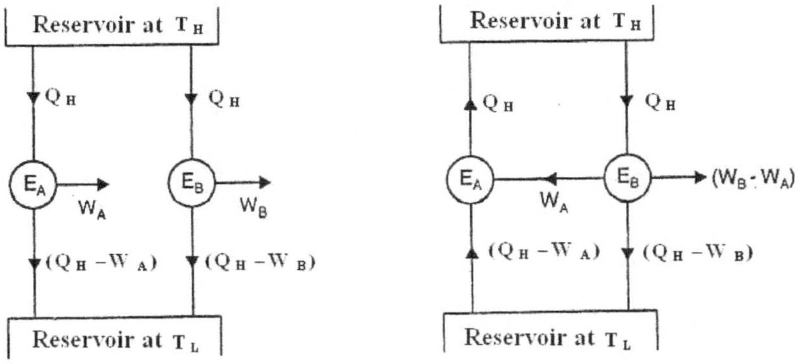

(a) (b)

Fig.3.6:Proof ofCarnotTheorem

Letitbepresumedthattheirreversibleengine$E_B$ismoreefficientthanthereversible engineE_A. Then $\frac{W_B}{Q_H} > \frac{W_A}{Q_H}$; $W_B > W_A$ and $(Q_H - W_A) > (Q_H - W_B)$ i.e.,workoutput fromirreversibleengineismorethanthatfromreversibleengine.

Let the reversible engine $E_A$now be made to operate as a refrigerator or heat pump; the irreversible engine continues to operate as an engine.

Since engine $E_A$is reversible, the magnitudes of heat and work interactions will remain the same but their direction will be reversed.The work required to drive the refrigerator can be withdrawn from the irreversible engine by having a direct coupling between the two.Fig (b) shows the work and heat interactions for the composite system constituted by the reversible engine (now operating as refrigerator) and the irreversible engine.The net effect is

- No net interaction with the high temperature heat reservoir.It supplies and recovers back the same amount of heat.
- The composite system withdraws $(Q_H - W_A) - (Q_H - W_B) = (W_B - W_A)$ units of heat from the low temperature reservoir and converts that into equivalent amount of work output.

Thecombinationthusconstitutesaperpetualmotionofthesecondkindinviolationofthe second law.Obviously the assumption that the irreversible engine is more efficient than the reversible engine is wrong.Hence, an irreversible engine cannot have efficiency higher than that from a reversible engine operating between the same thermal reservoirs.

CorollariesofCarnot'sTheorem:

Corollary1: All reversible engines operating between the two given thermal reservoirs, with fixed temperature, have the same efficiency.

Corollary2: The efficiency of any reversible heat engine operating between two reservoirs is independent of the nature of working fluid and depends only on the temperature of the reservoirs.

The Thermodynamic Temperature Scale

- Theconceptof thermodynamictemperature scale maybedeveloped withthehelp of figure 3.7, which shows three reservoirs and three engines that operate on the Carnot cycle.

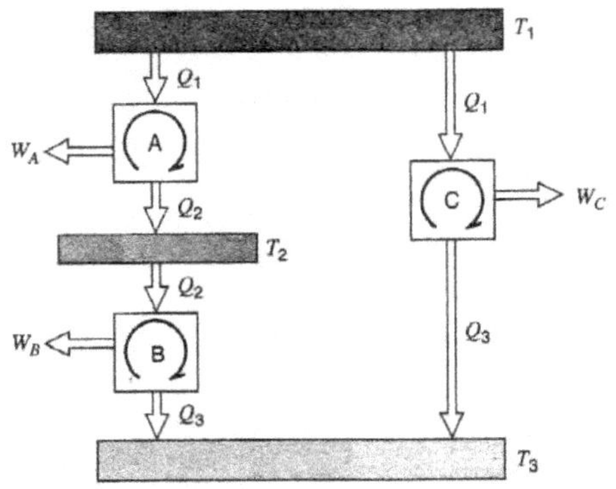

Fig.3.7:ArrangementofheatenginestodemonstratetheThermodynamicTemperature scale

- $T_1$is the highest temperature, $T_3$is the lowest temperature, and $T_2$is an intermediate temperature, and the engines operate between the various reservoirs as indicated.$Q_1$is the same for both A and C and, since we are dealing with reversible cycles, $Q_3$is the same for B and C.Since the efficiency of a Carnot cycle is a function only of the temperature, we can write

$$\eta_{thermal} = 1 - \frac{Q_L}{Q_H} = 1 - \frac{T_L}{T_H} = 1 - \psi(T_L, T_H)$$

Where ψ designates a functional relation. Let us apply this functional relation to the three Carnot Cycles shown in figure 3.7.

$$\frac{Q_1}{Q_2} = 1 - \psi(T_1, T_2); \quad \frac{Q_2}{Q_3} = 1 - \psi(T_2, T_3); \quad \frac{Q_1}{Q_3} = 1 - \psi(T_1, T_3)$$

Since $\frac{Q_1}{Q_3} = \frac{Q_1 Q_2}{Q_2 Q_3}$

It follows that $\psi(T_1, T_3) = \psi(T_1, T_2) \times \psi(T_2, T_3)$

Note that the left side is a function of T_1 and T_3 (and not of T_2), and therefore the right side of this equation must also be a function of T_1 and T_3 (and not of T_2). From this fact we can conclude that the form of the function ψ must be such that

$$\psi(T_1, T_2) = \frac{f(T_1)}{f(T_2)}; \quad \psi(T_2, T_3) = \frac{f(T_2)}{f(T_3)}$$

For in this way $f(T_2)$ will cancel from the product of $\psi(T_1, T_2) \times \psi(T_2, T_3)$. Therefore, we conclude that $\frac{Q_1}{Q_3} = \psi(T_1, T_3) = \frac{f(T_1)}{f(T_3)}$

In general terms, $\frac{Q_H}{Q_L} = \frac{f(T_H)}{f(T_L)}$

Suppose we had a heat engine operating on the Carnot cycle that received heat at the temperature of the steam point and rejected heat at the temperature of the ice point. The efficiency of such an engine could be measured to be 26.8%,

$$\eta_{th} = 1 - \frac{T_L}{T_H} = 1 - \frac{T_{icepoint}}{T_{steampoint}} = 0.2680$$

$$\frac{T_{icepoint}}{T_{steampoint}}=0.7320$$

This gives us one equation concerning the two unknowns T_H and T_L. The second equation comes from the difference between the steam point and ice point.

$$T_{steampoint}-T_{icepoints}=100$$

Solving these two equations simultaneously, we find

$$T_{steampoint}=373.15K \text{ and } T_{icepoint}=273.15K$$

It follows that, T(°C) + 273.15 = T(K)

The Inequality of Clausius

- The inequality of Clausius is expressed as $\oint\frac{\delta Q}{T}\leq 0$
- The inequality of Clausius is a corollary or a consequence of the second law of thermodynamics. It will be demonstrated to be valid for all possible cycles, including both reversible and irreversible heat engines and refrigerators.

Consider first a reversible (Carnot) heat engine cycle operating between reservoirs at temperatures T_H and T_L, as shown in Fig. 3.8. For this cycle, the cyclic integral of the heat transfer, $\oint \delta Q$, is greater than zero.

$$\oint\delta Q=Q_H-Q_L>0$$

Since T_H and T_L are constant, from the definition of the absolute temperature scale and from the fact this is a reversible cycle, it follows that

$$\oint\frac{\delta Q}{T}=\frac{Q_H}{T_H}-\frac{Q_L}{T_L}=0$$

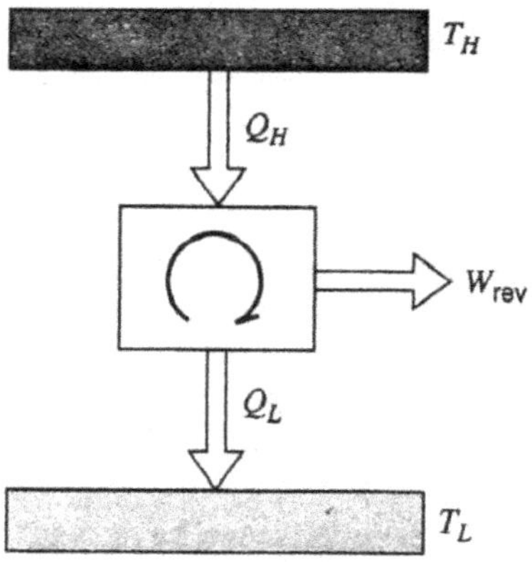

Fig.3.8: Reversible heat engine cycle for demonstration of the inequality of Clausius

If $\oint\delta Q$, the cyclic integral of δQ, approaches zero (by making T_H approach T_L) and the cycle remains reversible, the cyclic integral of $\delta Q/T$ remains zero. Thus, we conclude that for all reversible heat engine cycles

$$\oint\frac{\delta Q}{T}=0$$

Now consider an irreversible cyclic het engine operating between the same T_H and T_L as the reversible engine of Fig. 8.1 and receiving the same quantity of heat Q_H. Comparing the irreversible cycle with the reversible one, we conclude from the second law that

$$W_{irr}<W_{rev}$$

Since $Q_H - Q_L = W$ for both the reversible and irreversible cycles, we conclude that $(Q_H - Q_L)_{irr} < (Q_H - Q_L)_{rev}$ and therefore, $(Q_L)_{irr} > (Q_L)_{rev}$

Consequently, for the irreversible cyclic engine,

$$=(Q_H - Q_L)_{irr} > 0$$

$$\oint \frac{\delta Q}{T} = \frac{Q_H}{T_H} - \frac{(Q_L)_{irr}}{T_L} < 0$$

Thus, we conclude that for all irreversible heat engine cycles

$$\oint \frac{\delta Q}{T} < 0$$

Therefore, in general, for any heat engine or refrigerator $\oint \frac{\delta Q}{T} \leq 0$

The Significance of the inequality of Clausius may be illustrated by considering the simple steam power plant cycle shown in Fig.3.9.

Heat is transferred in two places, the boiler and the condenser. Therefore,

$$\oint \frac{\delta Q}{T} = \int \left(\frac{\delta Q}{T}\right)_{boiler} + \int \left(\frac{\delta Q}{T}\right)_{condenser}$$

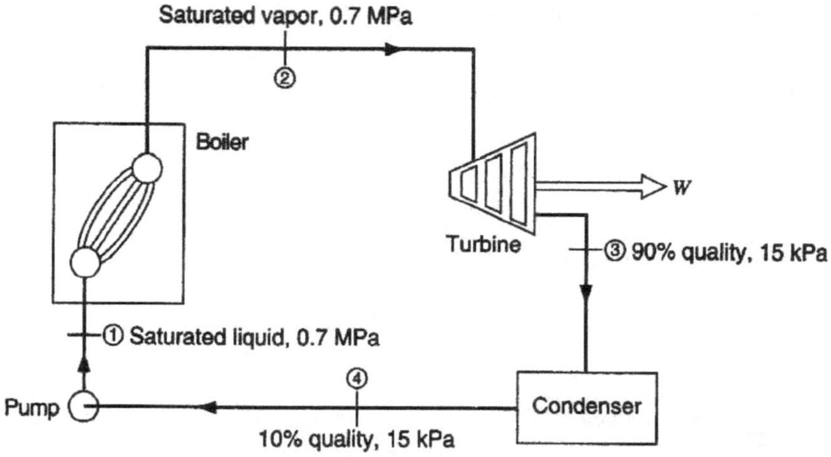

Fig.3.9: A simple Steam Power Plant that demonstrates the Inequality of Clausius

Since the temperature remains constant in both the boiler and condenser, this may be integrated as follows:

$$\oint \frac{\delta Q}{T} = \frac{1}{T_1}\int_1^2 \delta Q + \frac{1}{T_3}\int_3^4 \delta Q = \frac{Q_{12}}{T_1} + \frac{Q_{34}}{T_3}$$

Let us consider a 1 kg mass as the working fluid. We have then

$$q_{12} = h_2 - h_1 = 2066.3 kJ/kg, \qquad T_1 = 164.97^oC$$

$$q_{34} = h_4 - h_3 = 463.4 - 2361.8 = -1898.4 kJ/kg, \qquad T_3 = 53.97^oC$$

Therefore,

$$\oint \frac{\delta Q}{T} = \frac{2066.3}{164.97 + 273.15} - \frac{1898.4}{53.97 + 273.15} = -1.087 kJ/kg-K$$

Thus, this cycle satisfies the inequality of Clausius, which is equivalent to saying that itdoes not violate the second law of thermodynamics.

Entropy–APropertyofaSystem

ByapplyingEq.3.28andFig.3.10wecandemonstratethatthesecondlawofthermodynamics leads to a property of a system that we call entropy.

- *Entropyisameasureofmoleculardisorderlinessofasubstance.*

Let a closed system undergo a reversible process from state1 to state 2 along a pathA,and let the cycle be completed along path B, which is also reversible.

Becausethisisareversiblecycle,wecanwrite

$$\oint \frac{\delta Q}{T} = \int_1^2 \left(\frac{\delta Q}{T}\right)_A + \int_2^1 \left(\frac{\delta Q}{T}\right)_B = 0$$

Nowconsideranotherreversiblecycle,whichproceedsfirstalongpathCandisthen completed along path B.For this cycle we can write

$$\oint \frac{\delta Q}{T} = \int_1^2 \left(\frac{\delta Q}{T}\right)_C + \int_2^1 \left(\frac{\delta Q}{T}\right)_B = 0$$

Subtractingtheequation3.30fromtheequation3.29,weget

$$\int_1^2 \left(\frac{\delta Q}{T}\right)_A = \int_1^2 \left(\frac{\delta Q}{T}\right)_C$$

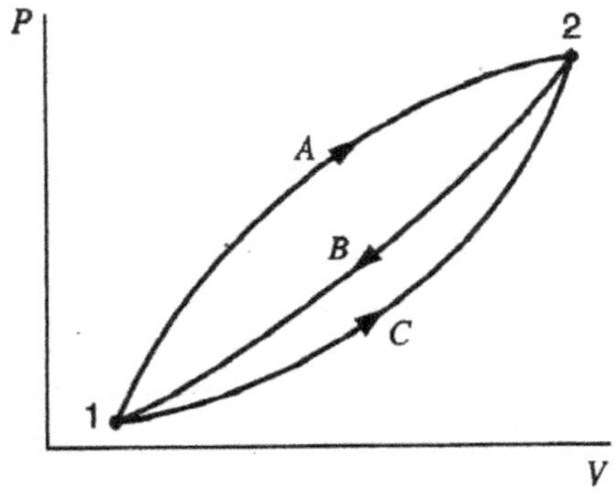

Fig.3.10:Tworeversiblecyclesdemonstratingthefactthatentropyisapropertyofa substance

Since the $\int \delta Q/T$ is the same for all reversible paths between states 1 and 2, we conclude that this quantityis independent of the path and it is a function of the end states only; it is therefore a property.This property is called entropy and is designated by 'S'.It follows that entropy may be defined as a property of a substance in accordance with the relation.

$$dS = \left(\frac{\delta Q}{T}\right)_{rev} \qquad \textbf{(Equation 3.32)}$$

- Entropy is an extensive property, and the entropy per unit mass is designated by 's'.The change in the entropy of a system as it undergoes a change of state may be found by integrating Eq.3.32.Thus,

$$S_2 - S_1 = \int_1^2 \left(\frac{\delta Q}{T}\right)_{rev}$$

- Toperformthisintegration,wemustknowtherelationbetweenTandQ.
- Sinceentropyisa property, the change in the entropyof asubstance ingoing from one state to another is the same for all processes, both reversible and irreversible, between these two states.
- Equation 3.33 enables us to calculate changes of entropy, but it tells us nothing about absolute values of entropy.
- From the third law of thermodynamics, which is based on observations of low-temperature chemical reactions, it is concluded that the entropy of all pure substances (in the appropriate structural form) can be assigned the absolute value of zero at the absolute zero of temperature.
- It also follows from the subject of statistical thermodynamics that all pure substances in the (hypothetical) ideal-gas state at absolute zero temperature have zero entropy.

Entropy Change of a control mass during a Reversible Process

Theentropychangeduringareversibleprocess1–2isgiven by

$$S_2 - S_1 = \int_1^2 \left(\frac{\delta Q}{T}\right)_{rev}$$

TherelationshipbetweenδQ andTcanbeobtainedfromthethermodynamicsproperty relations.

The Thermodynamic Property Relations

Thetwoimportantthermodynamicpropertyrelationsforacompressiblesubstancecanbe derived from the first law of thermodynamics.

$$\delta Q = dU + \delta W$$

Forareversibleprocessofsimplecompressiblesubstance,wecanwrite

$$\delta Q = TdS \text{and} \delta W = PdV$$

Substitutingtheserelationsintothefirstlawequation,weget

$$TdS = dU + PdV$$

Sinceenthalpyisdefinedas $= U + PV$,ondifferentiationweget

$$dH = dU + PdV + VdP$$

$$= \delta Q + VdP$$

$$= TdS + VdP \qquad \text{or}$$

$$TdS = dH - VdP$$

Theseequationscanalsobewrittenforaunitmass,

$$TdS = du + Pdv$$

$$TdS = dh - vdP$$

EntropyChangeofaControlMassduringanIrreversibleProcess

Consideracontrolmassthatundergoesthecyclesshowninfig.3.11.

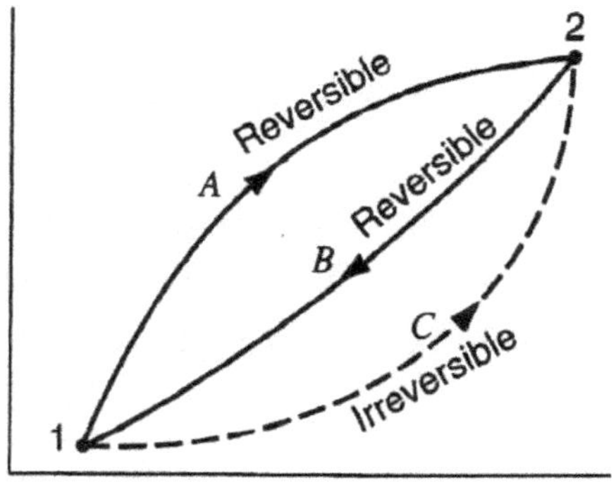

Fig.3.11:EntropyChangeofaControlMassduringanirreversibleprocess

Since the cycle made up of the reversible processes A and B is a reversible cycle, we can write

$$\oint \frac{\delta Q}{T} = \int_1^2 \left(\frac{\delta Q}{T}\right)_A + \int_2^1 \left(\frac{\delta Q}{T}\right)_B = 0$$

$\int_1^2 \left(\frac{\delta Q}{T}\right)_C < 0$

ThecyclemadeoftheirreversibleprocessCandthereversibleprocessBisan irreversible cycle.Therefore, for this cycle the inequality of Clausius may be applied as

$$\oint \frac{\delta Q}{T} = \int_1^2 \left(\frac{\delta Q}{T}\right)_C + \int_2^1 \left(\frac{\delta Q}{T}\right)_B$$

Subtractingtheequation3.34fromtheequation3.33andrearranging,wehave

$$\int_1^2 \left(\frac{\delta Q}{T}\right)_A > \int_1^2 \left(\frac{\delta Q}{T}\right)_C$$

SincepathAisreversible,andsinceentropyisaproperty,

$$\int_1^2 \left(\frac{\delta Q}{T}\right)_A = \int_1^2 dS_A = \int_1^2 dS_C$$

Therefore,

$$\int_1^2 dS_C > \int_1^2 \left(\frac{\delta Q}{T}\right)_C$$

AspathCwasarbitrary,thegeneralresultis

$$dS \geq \frac{\delta Q}{T}$$

or

$$S_2 - S_1 \geq \int_1^2 \frac{\delta Q}{T}$$

Intheseequationstheequalityholdsforareversibleprocessandtheinequalityforan irreversible process.

- Thus, If an amount of heat $\delta Q$is transferred to a control mass at temperature T ina reversible process, the change of entropy is given by the relation.

$$dS = \left(\frac{\delta Q}{T}\right)_{rev}$$

- If any irreversible effects occur while the amount of heat δQ is transferred to the control massattemperatureT, however,the changeof entropywillbe greater than that of the reversible process.We would then write

$$dS > \left(\frac{\delta Q}{T}\right)_{irr}$$

Entropy Generation

Theconclusionfromthepreviousconsiderationisthattheentropychangeforan irreversible process is larger than the change in a reversible process for the same $\delta Q$and T.Thiscanbewrittenoutinacommonformasanequality

$$dS = \frac{\delta Q}{T} + \delta S_{gen}$$

Providedthelasttermispositive,

$$\delta S_{gen} \geq 0$$

The amount of entropy, δS_{gen}, is the entropy generation in the process due to irreversibilities occurring inside the system.

- This internal generation can be caused by the processes such as friction, unrestrained expansions, and the internal transfer of energy (redistribution) over a finite temperature difference.
- Inadditiontothisinternalentropygeneration,externalirreversibilitiesarepossible by heat transfer over finite temperature differences as the δQ is transferred from a reservoir or by the mechanical transfer of work.
- We can generate but not destroy entropy.This is in contrast to energy which we can neither generate nor destroy.
- Since $\delta Q = 0$ for an adiabatic process, and the increase in entropy is always associated with the irreversibilities.
- The presence of irreversibilities will cause the actual work to be smaller than the reversible work.This means less work out in an expansion process and morework input in a compression process.

Principles of the increase of Entropy

Consider the process shown in Fig. 3.12 in which a quantity of heat $\delta Q$is transferred from the surroundings at temperature $T_0$to the control mass at temperature T.Let the work done during this process be δW.

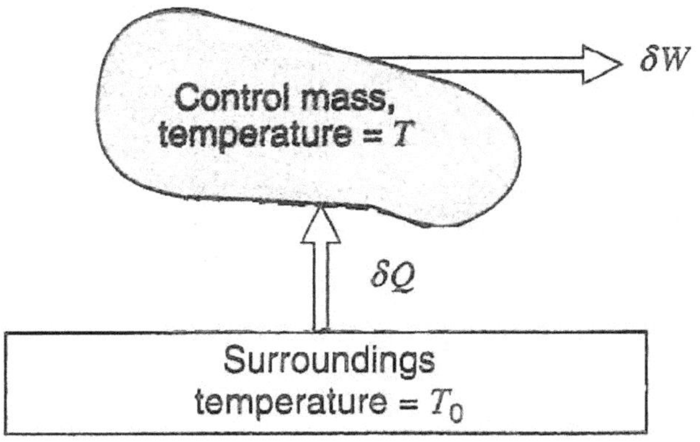

Fig.3.12:EntropyChangefortheControlMassPlusSurroundings

Forthisprocesswecanapplyequation3.38tothecontrolmassandwrite

$$\delta S_{c.m.} \geq \frac{\delta Q}{T}$$

ForthesurroundingsatT$_0$,δQisnegative,andweassumeareversibleheatextractionso

$$\delta S_{surr} = \frac{-\delta Q}{T_0}$$

Thetotalnetchangeofentropyistherefore

$$\delta S_{net} = \delta S_{c.m.} + \delta S_{surr} \geq \frac{\delta Q}{T} - \frac{\delta Q}{T_0}$$

$$\geq \delta Q\left(\frac{1}{T}-\frac{1}{T_0}\right)$$

If T > T_0, the heat transfer is from the control mass to the surroundings, and both δQ and the quantity {(1/T) – (1/T_0)} are negative, thus yielding the same result.

$$dS_{net}=dS_{c.m.}+dS_{surr}\geq 0$$

Thenetentropychangecouldalsobetermedthetotalentropygeneration:

$$dS_{net}=dS_{c.m.}+dS_{surr}=\sum \delta W_{gen}\geq 0$$

where the equality holds for reversible processes and the inequality for irreversible processes.

- This is a very important equation, not only for thermodynamics but also for philosophical thought.This equation is referred to as the principle of the increase of entropy.
- The great significance is that the only processes that can take place are those in whichthenetchangeinentropyofthecontrolmassplusitssurroundingsincreases (or in the limit, remain constant).The reverse process, in which both the control mass and surroundings are returned to their original state, can never be made to occur.
- Thus, the principle of the increase of entropy can be considered a quantitative general statement of the second law and applies to the combustion of fuel in our automobile engines, the cooling of our coffee, and the processes that take place in our body.

Entropy Change of a Solid or Liquid

Writingthefirstthermodynamicpropertyrelation,

$$Tds=du+Pdv$$

Sincechangeinspecificvolumeforasolidorliquidisverysmall,

$$ds\approx\frac{du}{T}\approx c\frac{dT}{T}$$

Formanyprocessesinvolvingasolidorliquid,wemayassumethatthespecificheat remains constant, in which case equation 3.47 can be integrated.The result is

$$s_2-s_1=cin\frac{T_2}{T_1}$$

Where,CisthespecificheatinJ/kg-K.

EntropyChangeofanIdealGas

Writingthefirstthermodynamicpropertyrelation,

$$Tds=du+Pdv$$

Foranideal gas,du=C_vdTand$\frac{P}{T}=\frac{R}{V}$

Therefore,ds=$c_v\frac{dT}{T}\pm R\frac{dv}{V}$

Uponintegration,wehave

$$s_2-s_1=\int_1^2 c_v\frac{dT}{T}+Rln\frac{v_2}{v_1}$$

Where, c_v is the specific heat at constant volume in J/kg –K. Similarly,

$$Tds = dh - vdP$$

For an ideal gas, $dh = c_p dT$ and $\frac{v}{T} = \frac{R}{P}$

Therefore, $ds = c_p \frac{dT}{T} - R \frac{dP}{P}$

Upon integration, we have

$$s_2 - s_1 = \int_1^2 c_p \frac{dT}{T} - R \ln \frac{P_2}{P_1}$$

Where, c_p is the specific heat at constant pressure in J/kg–K.

Entropy as a Rate Equation

$$\frac{dS}{\delta t} = \frac{1}{T}\frac{\delta Q}{\delta t} + \frac{\delta S_{gen}}{\delta t}$$

High and Low Grade Energy

High Grade Energy: High Grade Energy is the energy that can be completely transformed into shaft work without any loss and hence is fully utilizable. Examples are mechanical and electric work, water, wind and ideal power; kinetic energyof jets; animal and manual power.

Low Grade Energy: Low Grade Energy is the energy of which only a certain portioncan be converted into mechanical work.Examples are heat or thermal energy; heat from nuclear fission or fusion; heat from combustion of fuels such as coal, wood, oil, etc.

AvailableandUnavailableEnergy

- The portion of thermal energy input to a cyclic heat engine which gets convertedinto mechanical work is referred to as available energy.
- Theportionofthermalenergywhichisnotutilizableandisrejectedtothesink (surroundings) is called unavailable energy.
- Thetermsexergyandanergyaresynonymouswithavailableenergyandunavailable energy, respectively.Thus Energy = exergy+anergy.

Thefollowingtwocasesarisewhenconsideringavailableandunavailableportionsof heat energy

Case1: Heatiswithdrawnatconstanttemperature

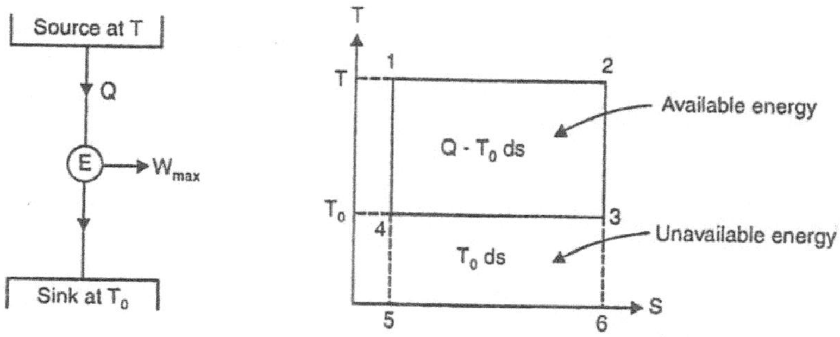

Fig.3.13:AvailableandUnavailableEnergy:HeatWithdrawnfromanInfinite Reservoir

Fig.3.13 represents a reversible engine that operates between a constant temperature reservoir at temperature T and a sink at temperature T_0.Corresponding to heat Q supplied by the reservoir, the available work $W_{max}$is given by

$$\eta=\frac{W_{max}}{Q}=\frac{T-T_0}{T}$$

Therefore,

$$W_{max}=\text{Availableenergy}=Q\left[\frac{T-T_0}{T}\right]=Q\left[1-\frac{T_0}{T}\right]=Q-T_0\frac{Q}{T}=Q-T_0\,ds$$

$$\text{Unavailableenergy}=T_0 ds$$

Where ds represents the change of entropy of the system during the process of heat supply Q.

Case2: Heatiswithdrawnatvaryingtemperature.

In case of a finite reservoir, the temperature changes as heat is withdrawn from it (Fig. 3.14), and as such the supply of heat to the engine is at varying temperature.The analysis is then made by breaking the process into a series of infinitesimal Carnot cycles each supplying δQ of heat at the temperature T (different for each cycle) and rejecting heat at the constanttemperatureT_0.Maximumamountof work (available energy) then equals

Fig.3.14:AvailableandUnavailableEnergy: Heat Supply at varying Temperature

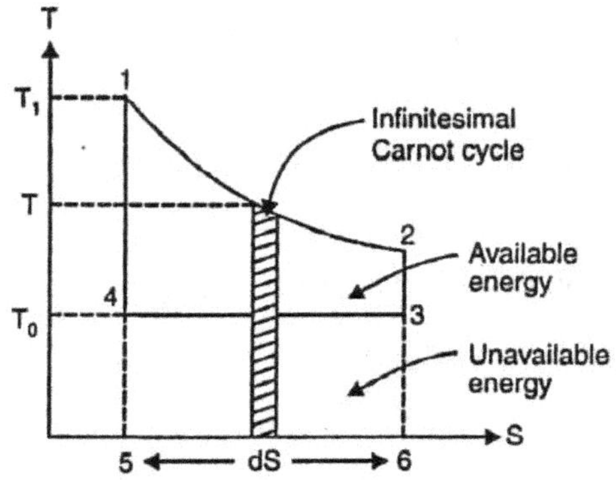

$$W_{max}=\int[1-\frac{T_0}{T}]\delta Q$$
$$=\int\delta Q-\int T_0\frac{\delta Q}{T}$$
$$=\int\delta Q-T_0\int\frac{\delta Q}{T}=Q-T_0\,ds$$

It is to be seen that expressions for both the available and unavailable parts are identical in the two cases.

LossofAvailableEnergyduetoHeatTransferthroughaFiniteTemperatureDifference

Consider a certain quantity of heat Q transferred from a system at constant temperature T_1 to another system at constant temperature T_2 ($T_1>T_2$) as shown in Fig. 3.15.

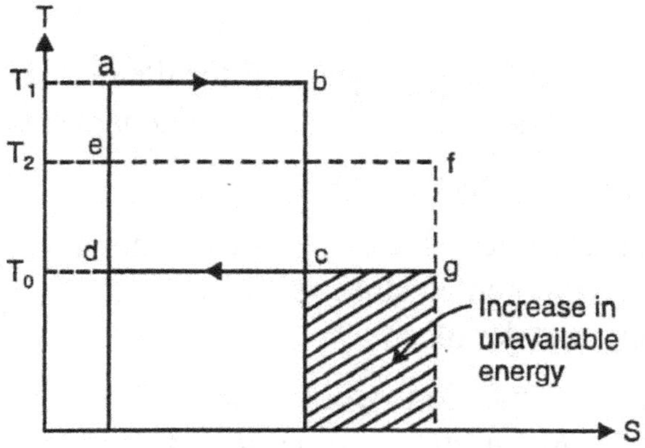

Fig.3.15: DecreaseinAvailableEnergyduetoHeatTransferthroughaFiniteTemperature Difference

Before heat is transferred, the energy Q is available at T_1 and the ambient temperature is T_0.

Therefore,Initialavailableenergy, $(AE)_1 = Q[1-\frac{T_0}{T_1}]$

After heat transfer,theenergyQisavailableat T_2 and againtheambient temperatureis T_0.

Therefore,Finalavailableenergy, $(AE)_2 = Q[1-\frac{T_0}{T_2}]$

Changeinavailable energy $=(AE)_1-(AE)_2=Q[1-\frac{T_0}{T_1}]-Q[1-\frac{T_0}{T_2}]$

$$=T_0[\frac{-Q}{T_1}+\frac{Q}{T_2}]=T_0(dS_1+dS_2)=T_0(dS)_{net}$$

Where $dS_1=-\frac{Q}{T_1}$, $dS_2=\frac{Q}{T_2}$ and $(dS)_{net}$ isthenetchangeintheentropyofthe combinationtothetwointeractingsystems.Thisnetentropychangeiscalledtheentropy changeofuniverseorentropyproduction.

Sincetheheattransferhasbeenthroughafinitetemperaturedifference,theprocessis irreversible, i.e., $(dS)_{net}>0$ and hence there is loss or decrease of available energy.

Theaboveaspectsleadustoconcludethat:

- Wheneverheatistransferredthroughafinitetemperaturedifference,thereisalways a loss of available energy.
- Greater the temperature difference (T_1-T_2), the more net increase in entropy and, therefore, more is the loss of available energy.
- The available energy of a system at a higher temperature is more than at a lowertemperature, and decreases progressively as the temperature falls.

- The concept of available energy provides a useful measure of the quality ofenergy.Energy is said to be degraded each time it flows through a finite temperature difference.The second law may, therefore, be referred to as law of degradation of energy.

Availability

- Theworkpotentialofasystemrelativetoitsdeadstate,whichexchangesheat solely with the environment, is called the *availability* of the system at that state.
- When a system and its environment are in equilibrium with each other, the system is said to be in its dead state.
- Specifically,asysteminadeadstateisinthermalandmechanicalequilibrium with the environment at $T_0$and P_0.
- The numerical values of (T_0, P_0) recommended for the dead state are those of the standard atmosphere, namely, 298.15 K and 1.01325 bars (1 atm).

AvailabilityofNon-floworClosedSystem

Consider a piston-cylinder arrangement (closed system) in which the fluid at P_1, V_1, T_1, expands reversibly to the environmental state with parameters P_0, V_0, T_0.The following energy (work and het) interactions take place:

- The fluid expands and expansion work $W_{exp}$is obtained.From the principle of energyconservation, $\delta Q= \delta W + dU$,*we get*:$-Q=W_{exp}+ (U_0- U_1)$

  Theheatinteractionisnegativeasitleavesthesystem.

  Therefore $W_{exp}=(U_1-U_0)-Q$

- The heat Q rejected by the piston-cylinde assembly may be made to run a reversible hea engine.The output from the reversible engin equals

  $W_{eng}= Q\,[1-\frac{T_0}{T_1}]=Q-T_0(S_1-S_0)$ **(Equ.3.60)**

The sum total of expansion work $W_{exp}$and the engin work W_{eng}gives maximum work obtainable from th arrangement.

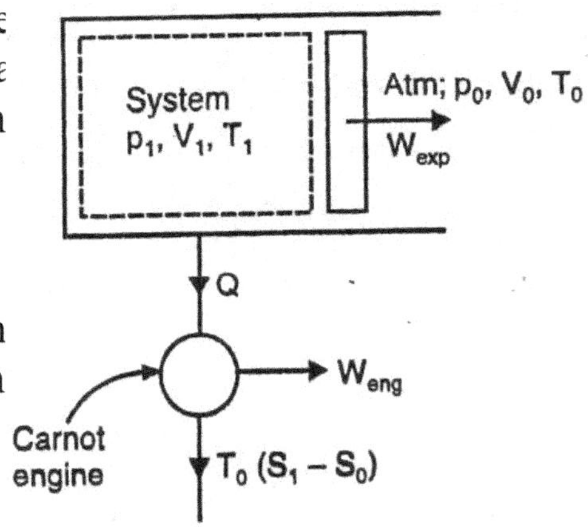

Fig.3.16:AvailabilityofaNon-Flow System

$$W_{max}=[(U_1-U_0)-Q]+[Q-T_0(S_1-S_0)]$$
$$=(U_1-U_0)-T_0(S_1-S_0)$$

The piston moving outwards has to spend a work in pushing the atmosphere against its own pressure.This work, which may be called as the surrounding work is simply dissipated, and as such is not useful.It is given by

$$W_{surr}=P_0(V_0-V_1)$$

Theenergyavailableforworktransferlesstheworkabsorbedin movingtheenvironment is called the useful work or net work.

Therefore,Maximumavailableusefulworkornetwork,

$$(W_{useful})_{max} = W_{max} - W_{surr}$$
$$= (U_1 - U_0) - T_0(S_1 - S_0) - P_0(V_0 - V_1)$$
$$= (U_1 + P_0V_1 - T_0S_1) - (U_0 + P_0V_0 - T_0S_0)$$
$$= A_1 - A_0$$

Where $A = (U + P_0V - T_0S)$ is known as non-flow availability function.It is a composite propertyof the system and surroundings as it consists of three extensive properties of the system (U, V and S) and two intensive properties of the surroundings ($P_0$and T_0).

When the system undergoes a change from state 1 to state 2 without reaching the dead state, then

$$(W_{useful})_{max} = W_{net} = (A_1 - A_0) - (A_2 - A_0) = A_1 - A_2$$

AvailabilityofSteadyFlowSystem

Consider a steady flow system and let it be assumed that the following fluid has the following properties and characteristics:

Internal energy U, specific volume V, specific enthalpy H, pressure P, velocity C and location Z.

Thepropertiesof thefluidwouldchangewhen flowingthroughthesystem.Letsubscript 1 indicate the properties of the system at inlet and subscript 0 be used to designate the fluid parameters at outlet corresponding to dead state.Further let Q units of heat be rejected by the system and let the system deliver W_{shaft}units of work.

$$U_1 + P_1V_1 + \frac{C_1^2}{2} + gZ_1 - Q = U_0 + P_0V_0 + \frac{C_0^2}{2} + gZ_0 + W_{shaft}$$

Neglectingpotentialandkineticenergychanges,

$$U_1 + P_1V_1 - Q = U_0 + P_0V_0 + W_{shaft}$$
$$H_1 - Q + H_0 + W_{shaft}$$

Therefore,Shaftwork $W_{shaft} = (H_1 - H_0) - Q$

TheheatQrejectedbythesystemmaybemadeto runa reversibleheat engine.The outputfromthis engine equals

$W_{eng} = Q\left[1 - \frac{T_0}{T_1}\right] = Q - T_0(S_1 - S_0)$ **(Equation3.67)**

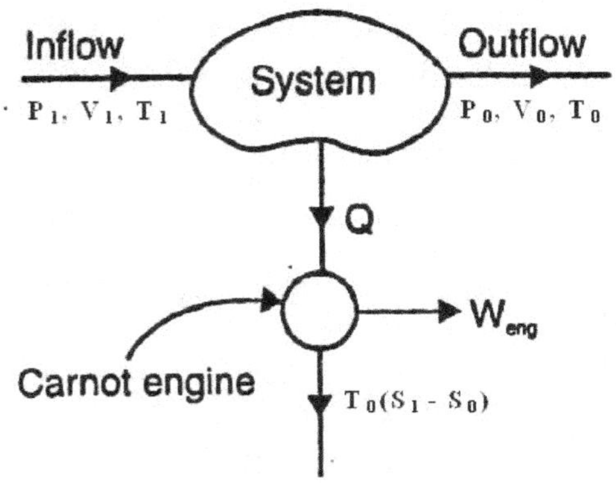

Therefore, Maximum available useful work or net work

$W_{net}=W_s+W_{eng}=(H_1-H_0)-Q+Q-T_0(S_1-S_0)$

$=(H_1-T_0S_1)-(H_0-T_0S_0)$

$=B_1-B_0$ **(Equation3.68)** **Fig.3.17:AvailabilityofaSteadyflowSystem**

Where $B = (H - T_0S)$ is known as the steady flow availability function.It is composite property of system and surroundings involving two extensive properties H and S of the system and one intensive property $T_0$of the surroundings.

When the system changes from state 1 to some intermediate state 2, the change in steady flow availability function is given by

$$= (B_1-B_0)- (B_2-B_0)=B_1-B_2$$

HelmholtzandGibb'sFunctions

Fora non-flowprocess,the maximumworkoutputfromthe system when $T_1= T_2= T_0$, is given by

$$W_{max}=(U_1-T_1S_1)-(U_2-T_2S_2)=A_1-A_2$$

- Theterm(U–TS)iscalledtheHelmholtzfunctionandisdefinedasthe differencebetweentheinternalenergyandtheproductoftemperatureandentropy.
- The maximum work of the process is equal to the decrease in Helmholtz function of the system.

In the case of flow process, the maximum work output from the system when$T_1= T_2= T_0$and neglecting kinetic and potential energies, is given by

$$W_{max}=(U_1+P_1V_1-T_1S_1)-(U_2+P_2V_2-T_2S_2)$$

$$=(H_1-T_1S_1)-(H_2-T_2S_2)=G_1-G_2$$

- The term (H – TS) is called Gibb's function and is defined as the difference between enthalpy and product of temperature and entropy.
- The changes of both Helmholtz and Gibb's functions are called free energy i.e., energy that is free to be converted into work.Further, both the Helmholtz and Gibb's functions establish a criterion for thermodynamic equilibrium.At equilibrium, these functions are at their minimum values.

Maxwell Relations

The Maxwell relations can be derived from the different forms of the thermodynamic property relations discussed earlier such as

$du=Tds-Pdv;dh=Tds+vdP;$

$A = U - TS$ (or) $a = u - Ts$ and $G = H - TS$ (or) $g = h - Ts$

Theserelationsareexactdifferentialsandare ofthegeneralform $dz = M\,dx + N\,dy$. For exact differentials, we have

$$\left(\frac{\partial M}{\partial y}\right)_x = \left(\frac{\partial N}{\partial x}\right)_y.$$

Thereforeusingthisrelationshipwecanderivethefollowingequations:

$$\left(\frac{\partial T}{\partial v}\right)_s = -\left(\frac{\partial P}{\partial s}\right)_v \quad ; \quad \left(\frac{\partial T}{\partial P}\right)_s = -\left(\frac{\partial v}{\partial s}\right)_P$$

$$\left(\frac{\partial P}{\partial T}\right)_v$$

$=-(\frac{\partial s}{\partial v})_T$ $\frac{\dot{}}{\partial T}_P$ $=-(\frac{\partial s}{\partial P})_T(\frac{\partial v}{})$ — —

These four equations are known as the Maxwell Relations for a simple compressiblefluid.

Third Law of Thermodynamics (Nernst Law)

At absolute zero temperature, the entropyof all homogeneous crystalline (condensed) substances in a state of equilibrium becomes zero. the molecules of a substance in solid phase continually oscillate, creating an uncertainty about their position. These oscillations, however, fade as the temperature is decreased, and the molecules supposedly become motionless at absolute zero.This represents a state of ultimate molecular order (and minimum energy). Therefore, *theentropyofapurecrystallinesubstanceatabsolutezerotemperatureiszero*sincethereisno uncertainty about the state of the molecules at that instant. This statement is known as the **third law of thermodynamics**. The third law of thermodynamics provides an absolute reference point for the determination of entropy. The entropy determined relative to this point is called **absolute entropy**, and it is extremely useful in the thermodynamic analysis of chemical reactions. Notice that the entropy of a substance that is not pure crystalline (such as a solid solution) is not zero at absolute zero temperature. This is because more than one molecular configuration exists for such substances, which introduces some uncertainty about themicroscopic state of the substance.

Unit – VI

Objectives:

- Tointroducethestudentstogaspowercycles,vapourpowercyclesand refrigeration cycles

Syllabus:

Gaspowercycles:Otto,diesel,dual combustioncycles,descriptionandrepresentation on P-v and T-s diagram, thermal efficiency, mean effective pressure on air standard basis-comparison of cycles, Brayton cycle

Vapour power cycles: Simple Rankine cycle, Refrigeration cycles: Bell Coleman cycle, vapour compression refrigeration system- performance evaluation.

Outcomes:

Studentswillbeable to

- Evaluate thermal efficiency of Otto, diesel, dual cycle and Brayton cycle andthey can represent on P-v and T-s diagrams
- Comparevariousgaspowercycles
- Understandvapourpowercycles
- Understandrefrigerationcycles
- Evaluateperformanceofrefrigerationcycles

- **Air-StandardAssumptions**

In our study of gas power cycles, we assume the working fluid is air, and the air undergoes a thermodynamic cycle even though the working fluid in the actual power system does not undergo a cycle.

Tosimplifytheanalysis,weapproximatethecycleswith thefollowingassumptions:

- Theaircontinuouslycirculatesinaclosedloopand alwaysbehavesasanidealgas.
- Alltheprocessesthatmakeupthecycleareinternallyreversible.
- Thecombustionprocessisreplacedbyaheat-additionprocessfromanexternal source.
- A heat rejection process that restores the working fluid to its initial state replaces the exhaust process.
- Thecold-air-standardassumptionsapplywhentheworkingfluidisairandhasconstant specific heat evaluated at room temperature (25°C or 77°F).

- **Terminology for Reciprocating Devices**

The following is some terminology we need to understand for reciprocating engines—typicallypiston-cylinder devices.Let's look at thefollowingfigures forthedefinitionsof top dead center (TDC), bottom dead center (BDC), stroke, bore, intake valve, exhaust valve, clearance volume, displacement volume, compression ratio, and mean effective pressure.

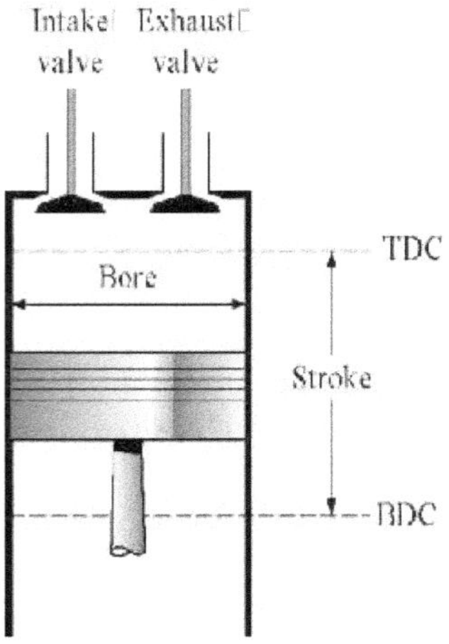

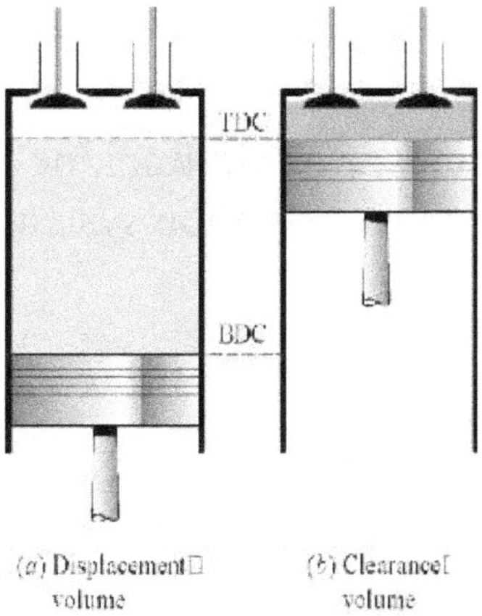

- Thecompressionratio*r*ofanengineistheratioofthemaximumvolumetotheminimum volume formed in the cylinder.

➢ The mean effective pressure **(MEP)** is a fictitious pressure that, if it operated on the piston during the entire power stroke, would produce the same amount of net work as that produced during the actual cycle.

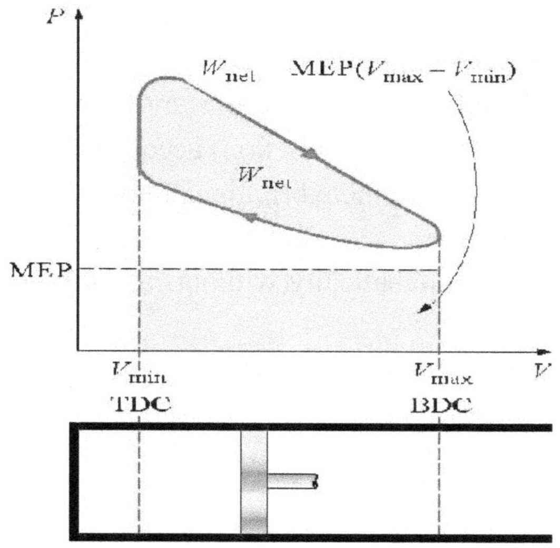

➢ Otto Cycle (Constant Volume Cycle):

This ideal heat engine cycle was proposed in 1862 by Bean de Rochas. In 1876 Dr. Otto designed an engine to operate on this cycle. The Otto engine immediately became so successful from a commercial stand point, that its name was affixed to the cycle used by it.

The ideal p - v and T-s diagrams of this cycle are shown in fig. In working out the air- standard efficiency of the cycle, the following assumptions are made:

(i) The working fluid (working substance) in the engine cylinder is air, and it behaves as a perfect gas, i.e., it obeys the gas laws and has constant specific heats.
(ii) The air is compressed adiabatically (without friction) according to law $pv^{\gamma} = C$
(iii)Theheatissuppliedtotheairatconstantvolumebybringingahotbodyincontact withtheend oftheenginecylinder.
(iv) Theairexpandsintheenginecylinderadiabatically(withoutfriction)duringthe expansion stroke.
(v) Theheatisrejected from theairat constantvolumebybringingacoldbodyin contact with the end of the engine cylinder.

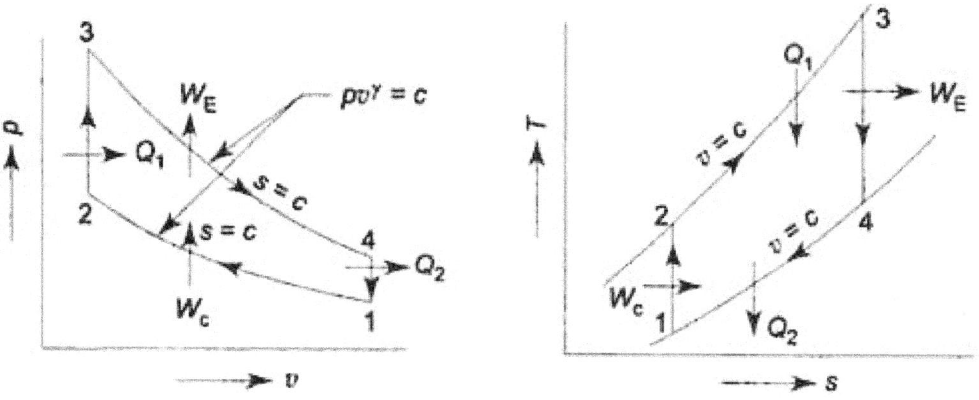

Process 1→ 2Isentropic compression
Process2→3Constantvolumeheataddition
Process 3 → 4 Isentropic expansion
Process4→1Constantvolumeheat rejection

Consider one kilogram of air in the engine cylinder at point (1). This air is compressed adiabaticallyto point (2), at which condition the hot bodyis placed in contact with the end of the cylinder. Heat is now supplied at constant volume, and temperature and pressure rise; this operation is represented by (2-3). The hot body is then removed and the air expands adiabatically to point (4). During this process, work is done on the piston. At point (4), the cold body is placed at the end of the cylinder. Heat is now rejected at constant volume, resultingindropoftemperatureandpressure.Thisoperationisrepresented by(4-1).Thecold bodyis then removed afterthe airis brought to its original state (condition). The cycleis thus completed. The cycle consists of two constant volume processes and two reversible adiabatic processes. The heat is supplied during constant volume process (2-3) and rejected during constant volume process (4-1). There is no exchange of heat during the two reversible adiabatic processes (1-2) and (3-4).

Theperformanceisoftenmeasuredintermsofthe cycle efficiency.

$$\eta_{th} = \frac{W_{net}}{Q_{in}}$$

ThermalEfficiencyoftheOtto cycle

$$\eta_{th} = \frac{W_{net}}{Q_{in}} = \frac{Q_{net}}{Q_{in}} = \frac{Q_{in} - Q_{out}}{Q_{in}} = 1 - \frac{Q_{out}}{Q_{in}}$$

Nowtofind Q_{in} and Q_{out}.

Applyfirst lawclosed system, V=constant.

Heat supplied during constant volume operation (2-3), Heat rejected during constant volume operation (4-1) is

$q_{in}=u_3-u_2=c_v(T_3-T_2)$ $q_{out}=u_4-u_1=c_v(T_4-T_1)$

$$Q_{net,23}=\Delta U_{23}$$

$$Q_{net,23}=Q_{in}=mC_v(T_3-T_2)$$

$$Q_{net,41}=\Delta U_{41}$$

$$Q_{net,41}=-Q_{out}=mC_v(T_1-T_4)$$

$$Q_{out}=-mC_v(T_1-T_4)=mC_v(T_4-T_1)$$

Thethermalefficiencybecomes

$$\eta_{th,Otto}=1-\frac{Q_{out}}{Q_{in}}$$

$$=1-\frac{mC_v(T_4-T_1)}{mC_v(T_3-T_2)}$$

$$\eta_{th,Otto}=1-\frac{(T_4-T_1)}{(T_3-T_2)}$$

$$=1-\frac{T_1(T_4/T_1-1)}{T_2(T_3/T_2-1)}$$

Recallprocesses1-2and3-4areisentropic,so

$$\frac{T_1}{T_2}=\left(\frac{v_2}{v_1}\right)^{\gamma-1}=\left(\frac{v_3}{v_4}\right)^{\gamma-1}=\frac{T_4}{T_3}$$

Since$V_3=V_2$and $V_4=V_1$

$$\frac{T_2}{T_1}=\frac{T_3}{T_4}$$

or

$$\frac{T_4}{T_1}=\frac{T_3}{T_2}$$

TheOttocycle efficiencybecomes

$$\eta_{th,Otto}=1-\frac{T_1}{T_2}$$

Sinceprocess1-2isisentropic,

$$\frac{T_2}{T_1}=\left(\frac{v_1}{v_2}\right)^{\gamma-1}=\left(\frac{1}{r}\right)^{\gamma-1}$$

Wherethecompressionratiois

$$r = \frac{v_{max}}{v_{min}} = \frac{v_1}{v_2}$$

$$\eta_{th,Otto} = 1 - \frac{1}{r^{k-1}}$$

- **Air-Standard Diesel Cycle (or constant pressure cycle):**
- Theair-standardDieselcycleistheidealcyclethatapproximatestheDieselcombustion engine

Process	Description
1-2	isentropic compression
2-3	Constantpressureheataddition
3-4	isentropic expansion
4-1	Constantvolumeheatrejection The

P-v and *T-s* diagrams are

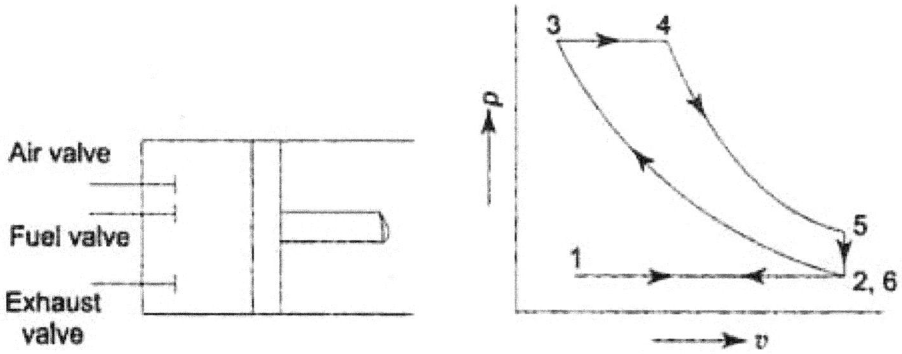

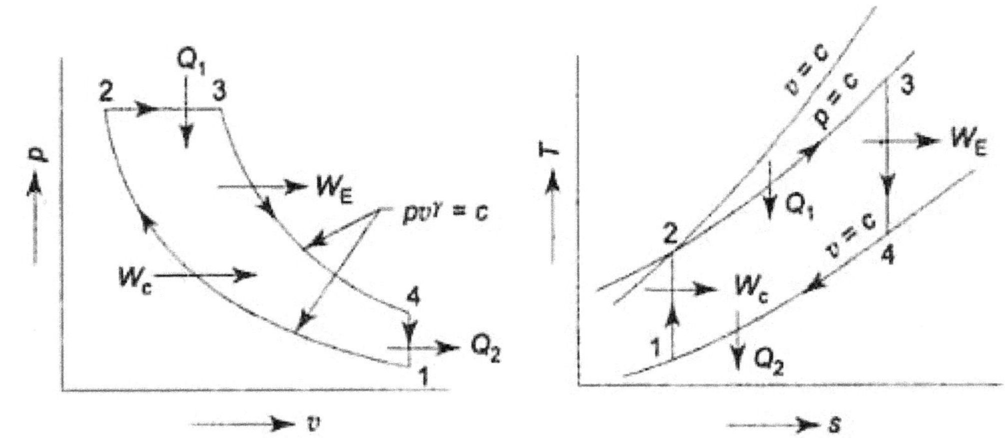

$Heat supplied = Q_1 = Q_{2-3} = mc_p(T_3 - T_2)$ Heat Rejected $= Q_2 = Q_{4-1} = mc_v(T_4 - T_1)$

Effieciency $= \eta = 1 - \frac{Q_2}{Q_1} = 1 - \frac{mc_v(T_4 - T_1)}{mc_p(T_3 - T_2)} = 1 - \frac{T_4 - T_1}{\gamma(T_3 - T_2)}$

Theefficiencymaybeexpressedtermsofanytwoofthe following.

$Compression Ratio = r_k = \frac{v_1}{v_2}$

$Expansion Ratio = r_e = \frac{v_4}{v_3}$

$Cut-off ratio = r_c = \frac{v_3}{v_2}$

$r_k = r_e . r_c$

Process3-4

$$\frac{T_4}{T_3} = \left(\frac{v_3}{v_4}\right)^{\gamma-1} = \frac{1}{r_e^{\gamma-1}}$$

$$T_4 = T_3\left(\frac{v_3}{v_4}\right)^{\gamma-1} = T_3\frac{r_c^{\gamma-1}}{r_k^{\gamma-1}}$$

Process2-3

$$\frac{T_2}{T_3} = \frac{p_2v_2}{p_3v_3} = \frac{v_2}{v_3} = \frac{1}{r_e}$$

$$T_2 = T_3\frac{1}{r_e}$$

Process1-2

$$\frac{T_1}{T_2}=\left(\frac{v_2}{v_1}\right)^{\gamma-1}=\frac{1}{r_k^{\gamma-1}}$$

$$T_1=T_2\frac{1}{r_k^{\gamma-1}}=\frac{T_3}{r_c}\frac{1}{r_k^{\gamma-1}}$$

BysubsistingT1,T2andT4 intheexpression ofefficiency

$$\eta=1-\frac{T_3\frac{r_c^{\gamma-1}}{r_k^{\gamma-1}}-\frac{T_3}{r_c}\frac{1}{r_k^{\gamma-1}}}{\gamma\left(T_3-T_3\frac{1}{r_c}\right)}$$

$$\eta_{diesel}=1-\frac{1}{\gamma}\frac{1}{r^{\gamma-1}}\cdot\frac{r_c^{\gamma-1}}{r_c-1}$$

- **Dual Cycle (mixed cycle/ limited pressure cycle):**

Process 1 → 2 Isentropic compressionProcess 2

→ 3Constant volume heat addition

Process3→4Constantpressureheataddition

Process 4 → 5 Isentropic expansion

Process5→1Constantvolumeheatrejection

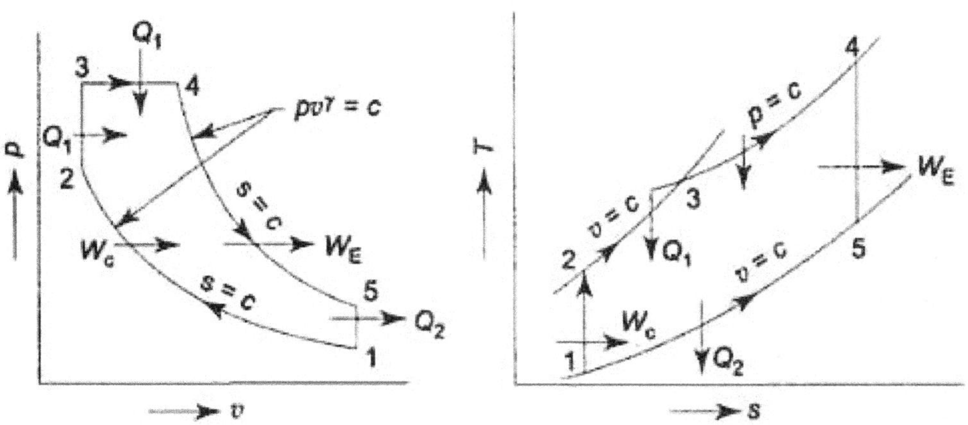

(a) (b)

ThermalEfficiency:

$$\text{where } r_c = \frac{v_3}{v_{2.5}} \text{ and } \alpha = \frac{P_3}{P_2}$$

Note,theOttocycle (r_c=1)andtheDieselcycle (a=1)arespecialcases:

$$\eta_{Otto} = 1 - \frac{1}{r^{k-1}} \qquad \eta_{Diesel\,const\,c_V} = 1 - \frac{1}{r^{k-1}}\left[\frac{1}{k}\frac{\left(r_c^{k}-1\right)}{\left(r_c-1\right)}\right]$$

Theuseofthe Dualcyclerequiresinformationabout either:

i) Thefractionsofconstantvolumeandconstantpressureheataddition (common assumption is to equally split the heat addition), or

ii) MaximumpressureP$_3$.

$$Q_1 = mc_v(T_3 - T_2) + mc_p(T_4 - T_3)$$

$$Q_2 = mc_v(T_5 - T_1)$$

HereQ_1=heatinput Q_2 =Out put

$$\eta = 1 - \frac{Q_2}{Q_1} = 1 - \frac{mc_v(T_5 - T_1)}{mc_v(T_3 - T_2) + mc_p(T_4 - T_3)} = 1 - \frac{T_5 - T_1}{T_3 - T_2 + \gamma(T_4 - T_3)}$$

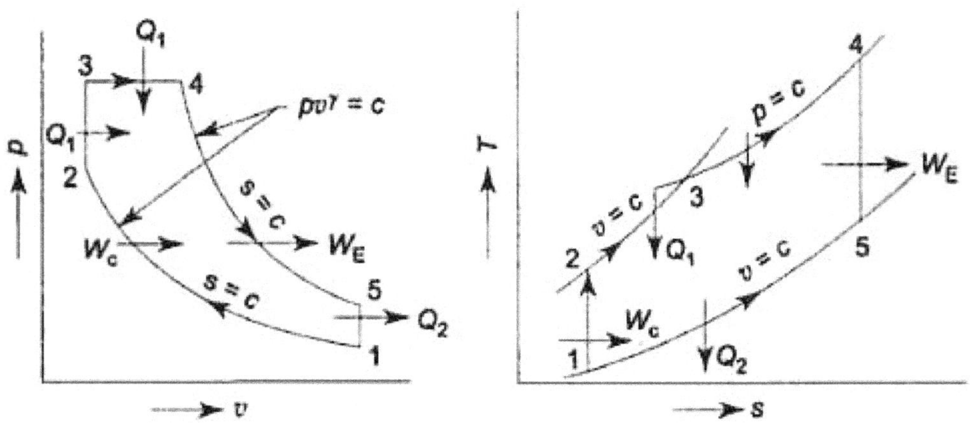

$$CompressionRatio = r_k = \frac{v_1}{v_2}$$

$$ExpansionRatio = r_e = \frac{v_4}{v_3}$$

$$constant-volume-pressure-ratio = r_p = \frac{p_3}{p_2}$$

$$r_k = r_c . r_e$$

$$r_e = \frac{r_k}{r_e}$$

$process 3-4$

$$r_c = \frac{v_4}{v_3} = \frac{T_4 p_3}{p_4 T_3} = \frac{T_4}{T_3}$$

$$T_3 = \frac{T_4}{r_c}$$

Process $2-3$

$$\frac{p_2 v_2}{T_2} = \frac{p_3 v_3}{T_3}$$

$$T_2 = T_3 \frac{p_2}{p_3} = \frac{T_4}{r_p r_c}$$

$process-1-2$

$$\frac{T_1}{T_2} = \left(\frac{v_2}{v_1}\right)^{\gamma-1} = \frac{1}{r_k^{\gamma-1}}$$

$$T_1 = \frac{T_4}{r_p r_c r_k^{\gamma-1}}$$

Process4-5

$$T_1 = \frac{T_4}{r_p r_c . r_k^{\gamma-1}}$$

$$\frac{T_5}{T_4} = \left(\frac{v_4}{v_5}\right)^{\gamma-1} = \frac{1}{r_e^{\gamma-1}}$$

$$T_5 = T_4 \frac{r_c^{\gamma-1}}{r_k^{\gamma-1}}$$

SubtitlingT1,T2,T3andT4values

$$\eta_{dual} = 1 - \frac{T_4 \frac{r_c^{\gamma-1}}{r_k^{\gamma-1}} - \frac{T_4}{r_p r_c r_k^{\gamma-1}}}{\left(\frac{T_4}{r_c} - \frac{T_4}{r_p r_c}\right) + \gamma\left(T_4 - \frac{T_4}{r_c}\right)} = 1 - \frac{1}{r_k^{\gamma-1}}\,\frac{r_p r_c^{\gamma} - 1}{r_p - 1 + \gamma r_p (r_c - 1)}$$

- **Comparison of cycles:**
  - ❖ Forthesameinletconditions P_1, V_1 and thesame compression ratio P_2/P_1:

Forthesameinitial conditions P_1, V_1 and thesamecompressionratio:

$\eta_{Otto} > \eta_{Dual} > \eta_{Diesel}$

- ❖ Forthesameinlet conditions P_1, V_1 and thesamepeak pressure P_3:

Forthesameinitialconditions P_1, V_1 andthesamepeakpressure P_3 (Actual design limitation in engines):

$\eta_{Diesel} > \eta_{Dual} > \eta_{otto}$

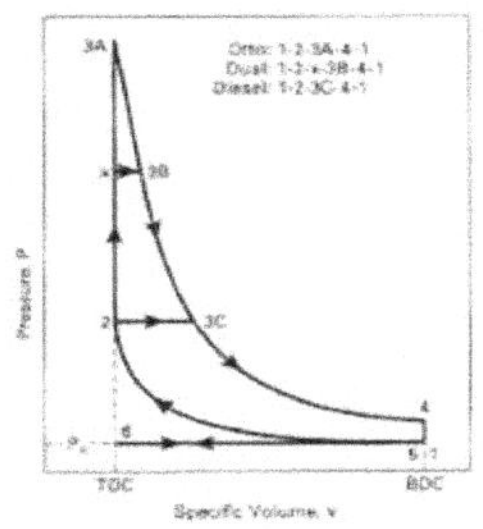

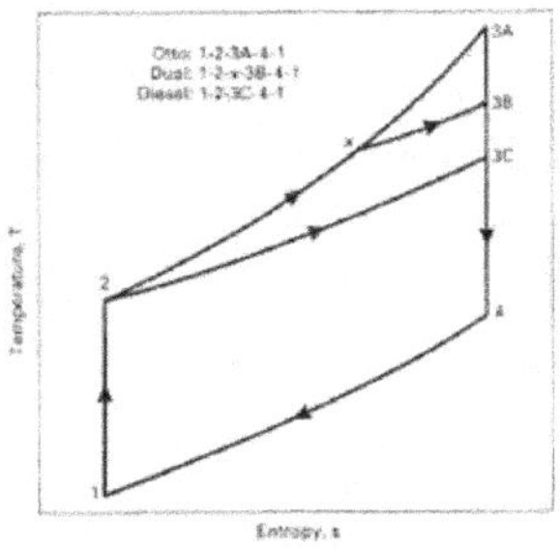

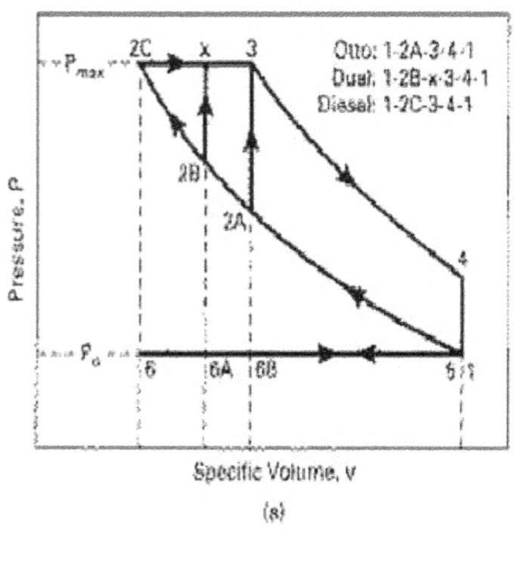

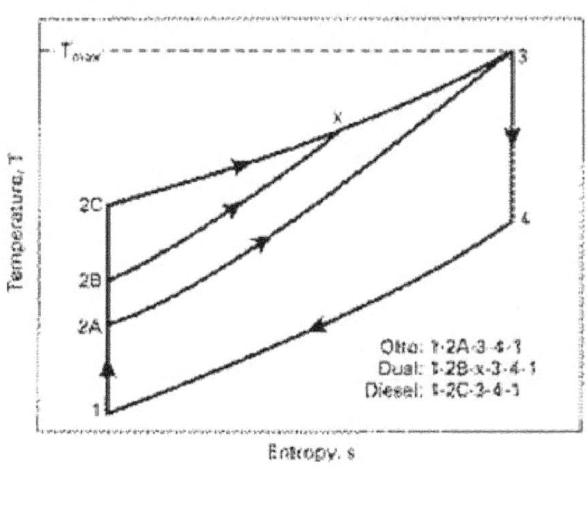

➢ Brayton Cycle (or Joule cycle)

The Brayton cycle is the air-standard ideal cycle approximation for the gas-turbine engine. This cycle differs from the Otto and Diesel cycles in that the processes making the cycle occur in open systems or control volumes.Therefore, an open system, steady-flow analysisis used to determine the heat transfer and work for the cycle.

We assume the working fluid is air and the specific heats are constant and will consider the cold-air-standard cycle.

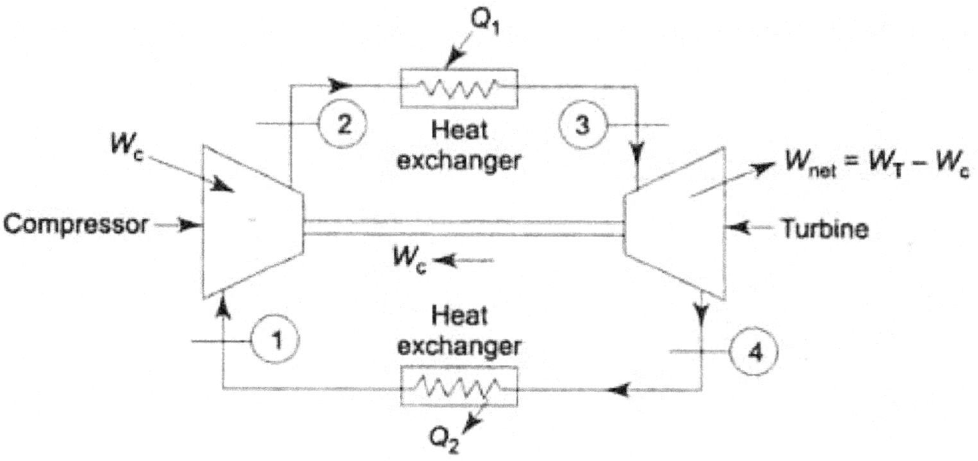

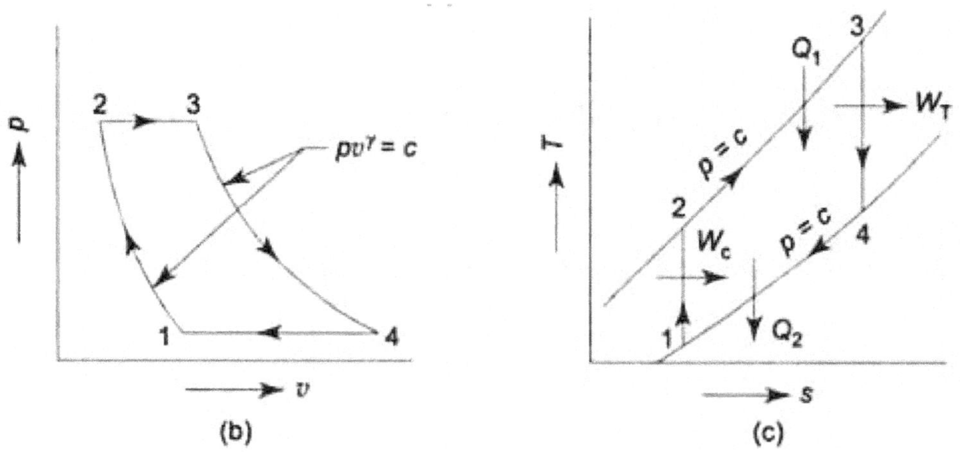

Theclosedcyclegas-turbineengine

The*T*-*s*and*P*-*v*diagramsforthe

ClosedBraytonCycle

Process Description

1-2 Isentropiccompression(inacompressor) 2-3

Constant pressure heat addition

3-4 Isentropicexpansion(inaturbine) 4-1

Constant pressure heat rejection

Thermal efficiency of the Brayton cycle

$$\eta_{th,Brayton} = \frac{W_{net}}{Q_{in}} = 1 - \frac{Q_{out}}{Q_{in}}$$

Nowtofind Q_{in} and Q_{out}.

Applythe conservation of energyto process 2-3 for P = constant (no work), steady-flow, and neglect changes in kinetic and potential energies.

$$\dot{E}_{in} = \dot{E}_{out}$$

$$\dot{m}_2 h_2 + \dot{Q}_{in} = \dot{m}_3 h_3$$

Theconservation ofmass gives

$$\dot{m}_{in} = \dot{m}_{out}$$

$$\dot{m}_2 = \dot{m}_3 = \dot{m}$$

Forconstant specificheats,theheat added perunit mass flowis

$$\dot{Q}_{in} = \dot{m}(h_3 - h_2)$$

$$\dot{Q}_{in} = \dot{m}C_p(T_3 - T_2)$$

$$q_{in} = \frac{\dot{Q}_{in}}{\dot{m}} = C_p(T_3 - T_2)$$

Theconservationofenergyforprocess4-1yieldsforconstantspecificheats

$$\dot{Q}_{out} = \dot{m}(h_4 - h_1)$$

$$\dot{Q}_{out} = \dot{m}C_p(T_4 - T_1)$$

$$q_{out} = \frac{\dot{Q}_{out}}{\dot{m}} = C_p(T_4 - T_1)$$

Thethermalefficiencybecomes

$$\eta_{th,Brayton} = 1 - \frac{\dot{Q}_{out}}{\dot{Q}_{in}} = 1 - \frac{q_{out}}{q_{in}}$$

$$= 1 - \frac{C_p(T_4 - T_1)}{C_p(T_3 - T_2)}$$

$$\boxed{\eta_{th,Brayton} = 1 - \frac{(T_4 - T_1)}{(T_3 - T_2)}}$$

$$= 1 - \frac{T_1(T_4/T_1 - 1)}{T_2(T_3/T_2 - 1)}$$

Recallprocesses1-2and3-4areisentropic,so

$$\frac{T_2}{T_1} = \left(\frac{P_2}{P_1}\right)^{\frac{\gamma-1}{\gamma}}$$

$$\frac{T_3}{T_4} = \left(\frac{P_3}{P_4}\right)^{\frac{\gamma-1}{\gamma}}$$

Since$P_3 = P_2$and $P_4 = P_1$,

$$\frac{T_2}{T_1} = \frac{T_3}{T_4}$$

or

$$\frac{T_4}{T_1} = \frac{T_3}{T_2}$$

TheBraytoncycleefficiencybecomes

$$\eta_{th,Brayton} = 1 - \frac{T_1}{T_2}$$

Sinceprocess1-2isisentropic,

$$\frac{T_2}{T_1} = \left(\frac{P_2}{P_1}\right)^{\frac{\gamma-1}{\gamma}} = r_p^{\frac{\gamma-1}{\gamma}}$$

$$\frac{T_1}{T_2} = \frac{1}{r_p^{\frac{(\gamma-1)}{\gamma}}}$$

Wherethepressureratiois$r_p = P_2/P_1$and

$$\boxed{\eta_{th,Brayton} = 1 - \frac{1}{r_p^{(\gamma-1)/\gamma}}}$$

SIMPLE RANKINE CYCLE

RankineCycle: Thesimplestwayofovercomingtheinherentpracticaldifficultiesofthe Carnot cycle without deviating too much from it is to keep the processes 1-2 and 2-3 of thelatterunchangedandtocontinuetheprocess3-4inthecondenseruntilallthevapourhas been converted into liquid water. Water is then pumped into the boiler upto the pressure correspondingto thestate1 and thecycleis completed. Such acycleis known as theRankine cycle. This theoretical cycle is free of all the practical limitations of the Carnot cycle.

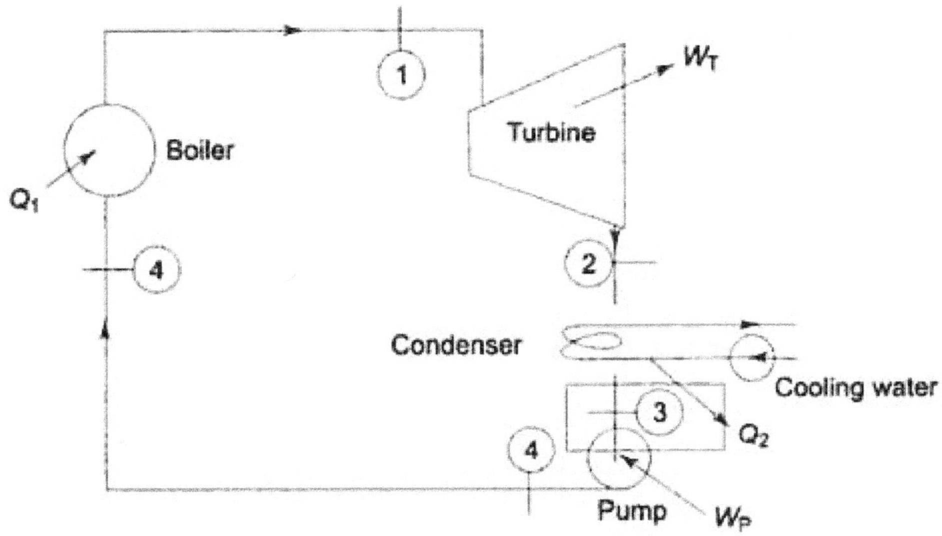

Figure (a) shows the schematic diagram for a simple steam power cycle which works on the principle of a Rankine cycle.

TheRankinecyclecomprisesthefollowingprocesses.

Process 1-2: Constant pressure heat transfer process in the boiler**Process 2-3:** Reversible adiabatic expansion process in the steam turbine **Process3-4:**Constantpressureheattransferprocessinthecondenserand **Process 4-1:** Reversible adiabatic compression process in the pump.

Figure(b)representsthe T-Sdiagramofthe cycle.

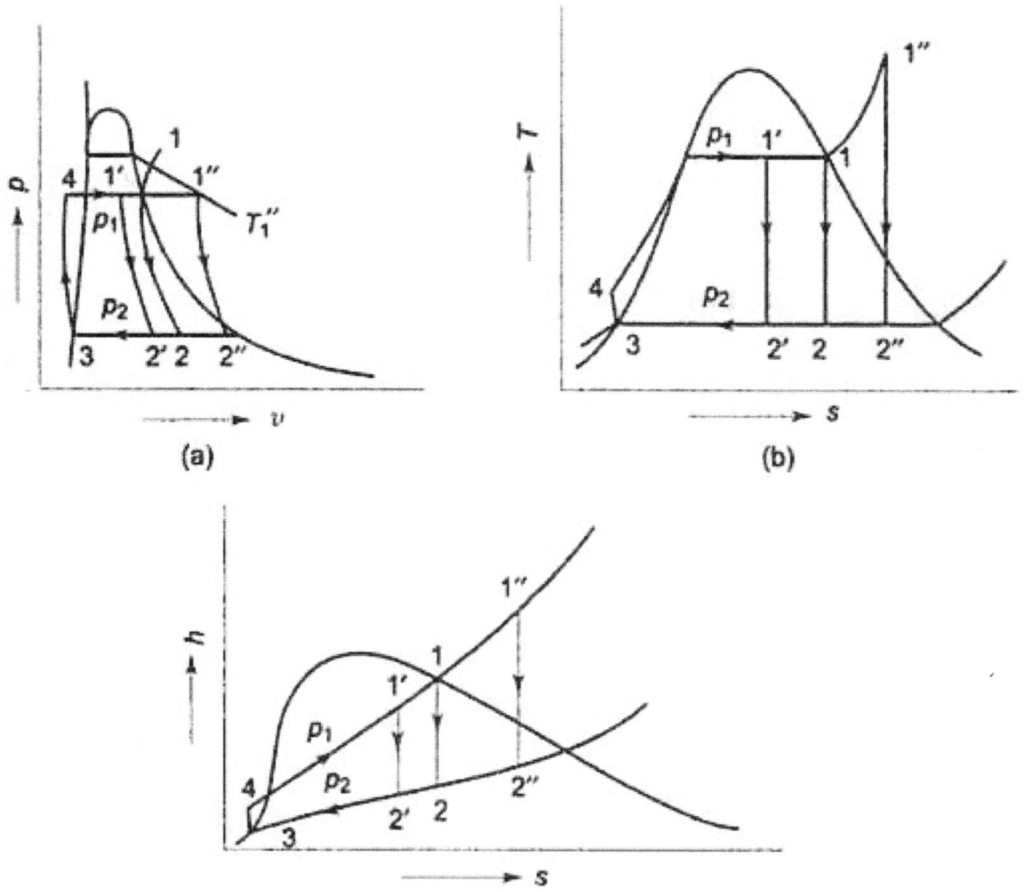

Thenumbersontheplotscorrespondtothenumbersontheschematicdiagram.Foranygiven pressure, the steam approaching the turbine may be dry saturated (state 2), wet (state 2^1) or superheated (state 2^{11}), but the fluid approaching the pump is, in each case, saturated liquid (state 4). Steam expands reversibly and adiabatically in the turbine from state 2 to state 3 (or 2^1 to 3^1 or 2^{11} to 3^{11}), the steam leaving the turbine condenses to water in the condenser reversibly at constant pressure from state 3 (or 3^1, or 3^{11}) to state 4. Also, the water is heated in the boiler to form steam reversiblyat constant pressure from state 1 to state 2 (or 2^1 or 2^{11})

ApplyingSFEEtoeachoftheprocessesonthebasisofunitmassoffluidandneglecting changes in KE & PE, the work and heat quantities can be evaluated

For1kgoffluid,theSFEEfortheboilerastheCV,gives, $h_4 + Q_1 =$

h_1i.e., $Q_1 = h_1 - h_4$ --- (1)

SFEEto turbine, $h_1 = W_T + h_2$i.e., $W_T = h_1 - h_2$---(2)

SFEEtocondenser,$h_2 = Q_2 + h_3$i.e.,$Q_L = h_2 - h_3$---(3) SFEE to pump,

$h_3 + W_P = h_4$ i.e., $W_P = h_4 - h_3$ --- (4)

TheefficiencyofRankinecycleis $\eta = \frac{W_{net}}{Q_1}$

TheefficiencyofRankinecycleis $\eta = \frac{W_{net}}{Q_1} = \frac{W_T - W_p}{Q_1} = \frac{(h_1 - h_2) - (h_4 - h_3)}{h_1 - h_4}$

VAPOUR COMPRESSION REFRIGERATION SYSTEM

In vapour compressionsystem,therefrigerantsusedare ammonia,carbon dioxide,freonsetc. the refrigerants alternately undergoes condensation and evaporation during the cycle. When refrigerant enters the evaporator it will be in liquid state and by absorbing latent heat it become vapours. Thus the C.O.P of this system is always much higher that air refrigeration systems.

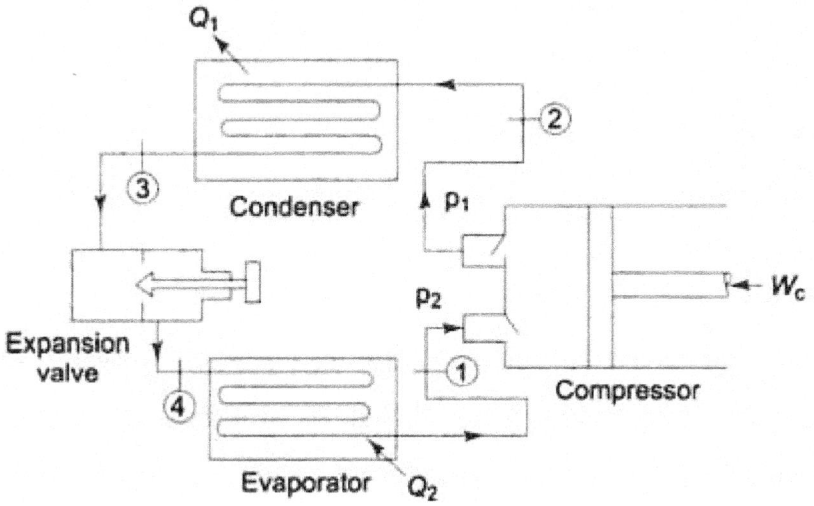

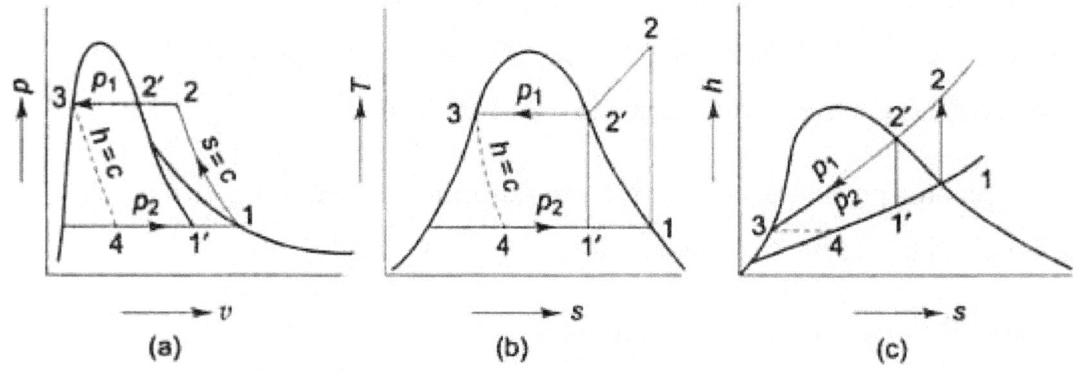

VAPOURCOMPRESSIONREFRIGERATIONSYSTEM

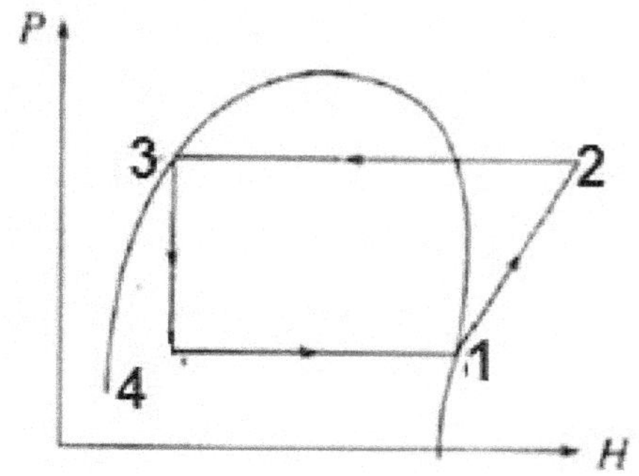

P-VDiagramv

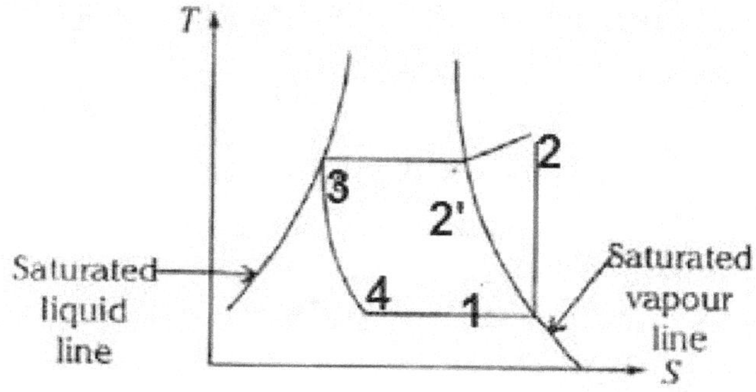

T-SDiagram

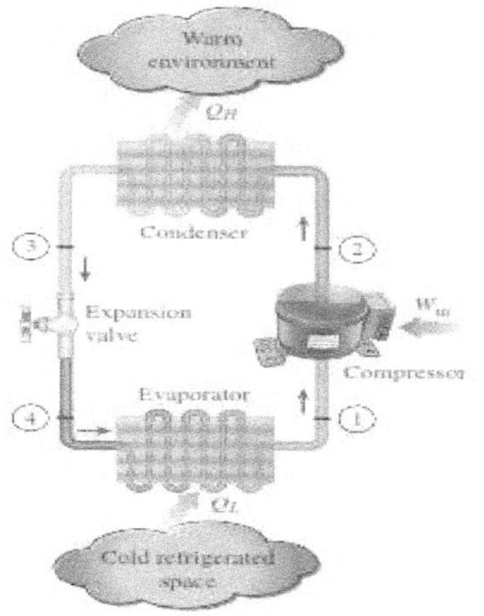

Thevariousprocessesare Process

1-2 .

Thevapourrefrigerantenteringthecompressoriscompressedtohighpressureand temperature in a isentropic manner.

Process2-3.

This high pressure and high temperature vapour then enters a condenser where the temperature of the vapour first drops to saturation temperature and subsequently the vapour refrigerants condenses to liquid state.

• **Process3-4.**

This liquidrefrigerant is collected in the liquidstorage tank and later on it is expanded tolow pressure and temperature by passing it through the

throttlevalve.Atpointdwehavelowtemperatureliquidrefrigerantwuitnsmall amount of

vapour.

• **Process4-1.**

This low temperature liquid then enters the evaporator where it absorbsheat from the spaceto be cooled namely the refrigerator and become vapour.

CompressorworkWc=h_2-h_1kJ/kg

Condenserwork =h_2-h_3 kJ/kg Expansion

value = h_3=h_4kJ/kg

$h_{f2}=h_{f1}+(x_4 h_{fg4})$

$$x_4=\frac{(hf_2-hf_1)}{hfg_4}$$

EvaporatorQ$_2$ =h_4-h_1kJ/kg

Refrigeranteffectisamountofheatremovedfromthesurroundingperunitmassflowrateof refrigerant

Coefficientofperformance $cop=\frac{Q_2}{w_c}=\frac{h_1-h_4}{h_2-h_1}$

Air Refrigeration System And Bell-Coleman Cycle Or Reversed Brayton Cycle

The components of the air refrigeration system are shown in Fig.(a). In this system, air is taken into the compressor from atmosphere and compressed. The hot compressed air iscooled in heat exchanger upto the atmospheric temperature (in ideal conditions). The cooled airis then expanded in an expander. Thetemperatureoftheair comingout from theexpander is below the atmospheric temperature due to isentropic expansion. The low temperature air coming out from the expander enters into the evaporator and absorbs the heat. The cycle is repeated again. The working of air refrigeration cycle is represented on p-v and T-s diagrams in Fig.(b) and (c).

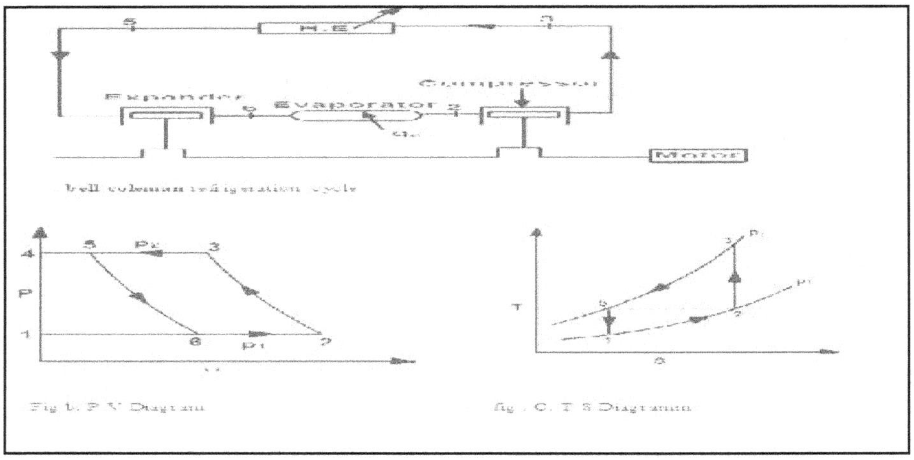

Process1-2representsthesuction ofairintothecompressor.

Process2-3represents theisentropiccompressionofairbythecompressor.

Process 3-5 represents the discharge of high pressure air from the compressor into the heat exchanger. The reduction in volume of air from v3 to v5 is due to the cooling of air in theheat exchanger. Process 5-6 represents the isentropic expansion of air in the expander.

Process6-2representstheabsorptionofheatfrom theevaporatoratconstantpressure

Assumptions:

1) Thecompressionandexpansionprocessesarereversibleadiabatic processes.

2) Thereisaperfectinter-coolingintheheat exchanger.

3) There arenopressurelossesinthesystem.

$$cop=\frac{\text{Net refrigeration effect Net}}{\text{work supplied}}$$

Netrefrigerationeffect Net

work supplied

Workdoneperkgofairfortheisentropiccompressionprocess2-3isgivenby, Wc= $C_p(T_3-T_2)$kJ/kg K

Workdevelopedperkgofairfortheisentropicexpansionprocess5-6isgivenby, $W_E= C_p(T_5-T_6)$ kJ/kg K

NetworkrequiredWnet=$(W_C-W_E)=(C_p(T_3-T_2))-(C_p(T_5-T_6))$kJ/kgK Net

$\frac{T_3}{T_2}=\left(\frac{p_3}{p_1}\right)^{\frac{\gamma-1}{\gamma}}$ $\frac{T_5}{T_6}=\left(\frac{p_5}{p_2}\right)^{\frac{\gamma-1}{\gamma}}$ $\frac{T_3}{T_2}=\frac{T_5}{T_6}$ $\frac{T_6}{T_5}=\frac{T_2}{T_5}$

n. T5= T2 or

refrigerating effect per kg of air is given by,

$R_{net}=C_p (T_2 -T_6)$kJ/kgK

$$cop=\frac{c_p(T_2-T_6)}{c_p(T_3-T_2)-(T_5-T_6)}$$

Performanceof Thesystem $cop=\frac{T_2}{T_3-T_2}$

Forperfectinter-cooling,therequiredconditionis$T_5=T_2$

$$cop=\frac{c_p(T_2-T_6)}{c_p(T_3-T_2)-(T_2-T_6)}$$

Prerequisites: Thermodynamics

CourseObjectives:

Theobjectiveofthiscourseistoprovideknowledge onworkingandperformanceofICengines, combustion process and compressors.

MODULEI: ActualCycles&I.C.Engines 12 Periods Actual Cycles and their Analysis: Introduction, comparison of air standard and actual cycles. time loss factor, heat loss factor, exhaust blow down, loss due to gas exchange process, volumetric efficiency. loss due to rubbing friction, actual and fuel - air cycles of CI Engines.

I.C. Engines: Classification - working principles, valve and port timing diagrams, air standard efficiency, air - fuel and actual cycles - engine systems - fuel, carburetor, fuel injection system, ignition, cooling and lubrication.

MODULEII: CombustioninS.I.Engines&C.I.Engines 13 Periods

Combustion in S.I. Engines : Normal Combustion and abnormal combustion – Importance of flame speed and effect of engine variables – Type of Abnormal combustion, pre- ignition and knocking – Fuel requirements and fuel rating, antiknock additives – combustion chamber – requirements, types.

CombustioninC.I.Engines:fourstagesofcombustion–delayperiodanditsimportance – effect of engine parameters – diesel knock – need for air movement, suction,compressionandcombustioninducedturbulence – openanddividedcombustionchambers and nozzles used –fuel requirements and fuel rating.

MODULEIII:I.C.EngineTestingandPerformance 13 Periods A: Testing and Performance - I: Terminologies, performance parameters - measurement of cylinder pressure, fuel consumption, air intake, exhaust gas composition, torque, brake power, dynamometer working & types.

B: Testing and Performance - II: Determination of brake power, frictional power, indicated power and mechanical efficiency – volumetric efficiency – MEP – Performance test – Heat balance test - Morse Test.

MODULEIV:ReciprocatingCompressor 13 Periods Introduction–Classification–positivedisplacementandrotordynamicmachinery – Power producing and power absorbing machines, fan, blower and compressor – positive displacement and dynamic types – reciprocating and rotary types.

Reciprocating Compressor: Principle of operation, work required, Isothermal efficiency volumetric efficiency and effect of clearance, stage compression, under cooling, saving of work, minimum work condition for stage compression.

MODULEV:Rotary&AxialFlowCompressors 13 Periods RotaryCompressors:RootsBlower,vanesealedcompressor,Lysholmcompressor – mechanical details and principle of working – efficiency considerations. Centrifugal compressors: Mechanical details and principle of operation – velocity and pressure variation. Energy transfer-impeller blade shape - losses, slip factor, power input factor, pressure coefficient and adiabatic coefficient – velocity diagrams – power.

Axial Flow Compressors: Mechanical details and principle of operation – velocitytriangles and energy transfer per stage degree of reaction, work done factor - isentropic efficiency - pressure rise calculations – Polytrophic efficiency.

TEXTBOOKS

➢ V.Ganesan,**"I.C.Engines"**,TataMcGrawHilleditions,4th edition,2012.

➢ SMYahya,**"Turbines,Pumps,Compressors"**,TataMcGrawHilleditions,4th edition, 2005.

3 R.K.Rajput,**"ThermalEngineering**",LakshmiPublications,10th Edition,2017

REFERENCES

(iii) Mathur&Sharma,**"ICEngines"**,DhanpathRai&Sons,4th edition,2010

(iv) Pulkrabek,**"EngineeringfundamentalsofICEngines"**,PearsonPHI,2nd edition,2004.

(v) Rudramoorthy,**"ThermalEngineering"**,TMHPublishers,3rd edition,2003.

(vi) B.Yadav,**"Thermodynamics&HeatEngines**",CentralBookDepot.,Allahabad, 6th edition, 2012.

(vii) B.SrinivasuluReddy,**"ThermalEngineeringdatabook"**,JKInternationalPub,2014.

E- RESOURCES

vi. https://www.journals.elsevier.com/applied-thermal-eng
vii. eds.yildiz.edu.tr/journal-of-thermal-engineering
viii. https://www.researchgate.net/... /1359-4311_Applied_Thermodynamics
ix. www.nptelvideos.com/mechanical/
x. https://www.btechguru.com/courses--nptel--mechanical

HEAT ENGINES

Module - IINTRODUCTION

Heat engine:

Aheatengineisadevicewhichtransformsthechemicalenergyofafuelintothermalenergy and uses this energy to produce mechanical work. It is classified into two types-

- Externalcombustionengine
- Internalcombustionengine

Externalcombustionengine:

Inthisengine,theproductsofcombustionofairandfueltransferheattoasecondfluid which is the working fluid of the cycle.

Examples:

*Inthesteam engine or a steam turbine plant, the heat ofcombustion isemployed to generate steam which is used in a piston engine (reciprocating type engine) or a turbine (rotary type engine) for useful work.

*In a closed cycle gas turbine, the heat of combustion in an external furnace is transferred to gas, usually air which the working fluid of the cycle.

Internalcombustionengine:

In this engine, the combustion of air and fuels take place inside the cylinder and are used as the direct motive force. It can be classified into the following types:

- Accordingtothebasicenginedesign-(a)Reciprocatingengine(Useofcylinderpiston arrangement), (b) Rotary engine (Use of turbine)
- Accordingtothetypeoffuelused-(a)Petrolengine,(b)dieselengine,(c)gasengine (CNG, LPG), (d) Alcohol engine (ethanol, methanol etc)
- Accordingtothenumberofstrokespercycle-(a)Fourstrokeand(b)Twostrokeengine
- Accordingtothemethodofignitingthe fuel-(a)Sparkignitionengine,(b)compression ignition engine and (c) hot spot ignition engine
- Accordingtotheworkingcycle-(a)Ottocycle(constantvolumecycle)engine,(b)diesel cycle(constantpressurecycle)engine,(c)dualcombustioncycle (semidieselcycle)engine.

- According to the fuel supply and mixture preparation- (a) Carburetted type (fuel supplied through the carburettor), (b) Injection type (fuel injected into inlet ports or inlet manifold,fuel injected into the cylinder just before ignition).

- Accordingtothenumberofcylinder-(a)Singlecylinderand(b)multi-cylinderengine

- Methodofcooling-watercooledoraircooled

- Speedoftheengine-Slowspeed,mediumspeedandhighspeedengine

- Cylinder arrangement-Vertical, horizontal, inline, V-type, radial, opposed cylinder or piston engines.

- Valve or port design and location- Overhead (I head), side valve (L head); in two stroke engines: cross scavenging, loop scavenging, uniflow scavenging.

- Method governing- Hit and miss governed engines, quantitatively governed engines and qualitatively governed engine

14. Application- Automotive engines for land transport, marine engines for propulsion of ships, aircraft engines for aircraft propulsion, industrial engines, prime movers for electrical generators.

<u>Comparisonbetweenexternalcombustionengineandinternalcombustionengine:</u>

Externalcombustionengine	Internalcombustionengine
*Combustionofair-fuelisoutsidetheengine cylinder (in a boiler)	➢ Combustionofair-fuelisinsidetheengine cylinder (in a boiler)
*Theenginesarerunningsmoothlyand silently due to outside combustion	➢ Verynoisyoperated engine
*Higherratioofweightandbulktooutput duetopresenceofauxiliaryapparatuslike boiler and condenser. Hence it is heavy and cumbersome.	➢ Itislightandcompactduetolowerratioof weight and bulk to output.
*Working pressure and temperature insidethe engine cylinder is low; hence ordinary alloysareusedforthemanufacture ofengine cylinder and its parts.	➢ Working pressure and temperature inside the engine cylinder is very much high; hence special alloys are used
*Itcanusecheaperfuelsincludingsolidfuels	*Highgradefuelsareusedwithproper filtration
*Lowerefficiencyabout15-20%	*Higherefficiencyabout 35-40%
➢ Higherrequirementofwaterfordissipation of energy through cooling system	*Lesserrequirementofwater
*Highstartingtorque	*ICenginesarenotself-starting

Main components of reciprocating IC engines:

Cylinder: It is the main part of the engine inside which piston reciprocates to and fro. It shouldhavehighstrengthtowithstandhighpressureabove50barandtemperatureabove2000 °C. The ordinary engine is made of cast iron and heavy duty engines are made of steel alloys or aluminum alloys. In the multi-cylinder engine, the cylinders are cast in one block known as cylinder block.

Cylinder head: The top end of the cylinder is covered bycylinder head over which inlet and exhaust valve, spark plug or injectors are mounted. A copper or asbestos gasket is provided between the engine cylinder and cylinder head to make an air tight joint.

Piston: Transmit the force exerted by the burning of charge to the connecting rod. Usually made of aluminium alloy which has good heat conducting property and greater strength at higher temperature.

Figure1showsthedifferentcomponentsofICengine.

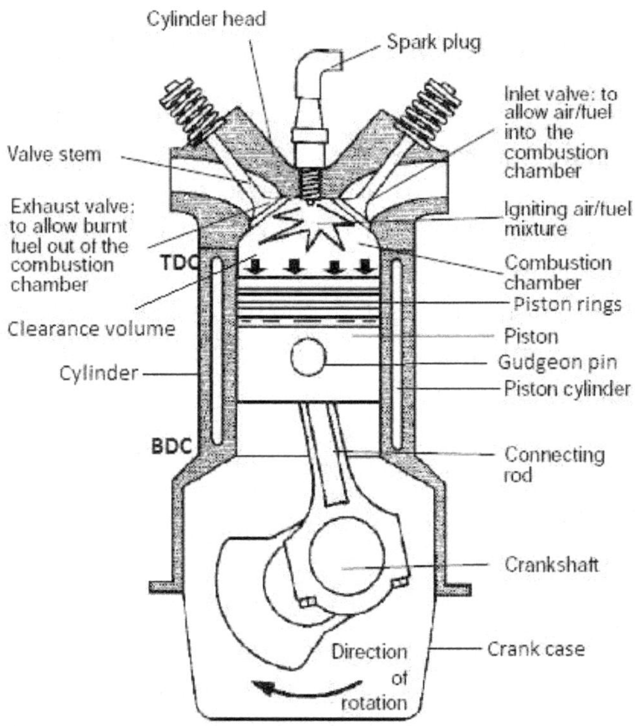

Fig.1.DifferentpartsofICengine

Piston rings: These are housed in the circumferential grooves provided on the outer surface ofthepistonandmadeofsteelalloyswhichretainelasticpropertiesevenathightemperature. 2 types of rings- compression and oil rings. Compression ring is upper ring of the piston which provides air tight seal to prevent leakage of the burnt gases into the lower portion. Oil ring is lower ring which provides effective seal to prevent leakage of the oil into the engine cylinder.

Connecting rod: It converts reciprocating motion of the piston into circular motion of the crank shaft, in the working stroke. The smaller end of the connecting rod is connected with the piston by gudgeon pin and bigger end of the connecting rod is connected with the crankwith crank pin. The special steel alloys or aluminium alloys are used for the manufacture of connecting rod.

Crankshaft: Itconvertsthereciprocatingmotionofthepistonintotherotarymotionwiththe help of connecting rod. The special steel alloys are used for the manufacturing of the crankshaft. It consists of eccentric portion called crank.

Crank case: It houses cylinder and crankshaft of the IC engine and also serves as sump for the lubricating oil.

Flywheel: It is big wheel mounted on the crankshaft, whose function is to maintain its speed constant. It is done by storing excess energy during the power stroke, which is returnedduring other stroke.

TerminologyusedinICengine:

➢ Cylinderbore(D):Thenominalinnerdiameteroftheworkingcylinder.

➢ Pistonarea(A):Theareaofcircleofdiameterequaltothecylinderbore.

➢ Stroke (L): The nominal distance through which a working piston moves between two successive reversals of its direction of motion.

➢ Dead centre: The position of the working piston and the moving parts which are mechanically connected to it at the moment when the direction of the piston motion is reversed (at either end point of the stroke).

d) Bottomdeadcentre(BDC):Deadcentrewhenthepistonisnearesttothecrankshaft.

e) Topdeadcentre(TDC):Deadcentrewhenthepositionisfarthestfromthe crankshaft.

➢ Displacement volume or swept volume (V_s): The nominal volume generated by the working piston when travelling from the one dead centre to next one and given as,

$$V_s=A\times L$$

➢ Clearance volume (V_c): the nominal volume of the space on the combustion side of the piston at the top dead centre.

➢ Cylindervolume(V):Totalvolumeofthecylinder.

$$V=V_s+V_c$$

➢ Compressionratio(r): —

Fourstrokeengine:

- Cycleofoperationcompletedinfourstrokesofthepistonortworevolutionofthe piston.

- Suction stroke (suction valve open, exhaust valve closed)-charge consisting of freshairmixedwiththefuelisdrawnintothecylinderduetothevacuumpressure created by the movement of the piston from TDC to BDC.
- Compression stroke (both valves closed)-fresh charge is compressed intoclearance volume by the return stroke of the piston and ignited by the spark for combustion. Hence pressure and temperature is increased due to the combustionof fuel
- Expansion stroke (both valves closed)-high pressure of the burnt gases force the piston towards BDC and hence power is obtained at the crankshaft.
- Exhaust stroke (exhaust valve open, suction valve closed)- burned gases expel out due to the movement of piston from BDC to TDC.

Figure2showthecycleofoperationoffourstroke engine.

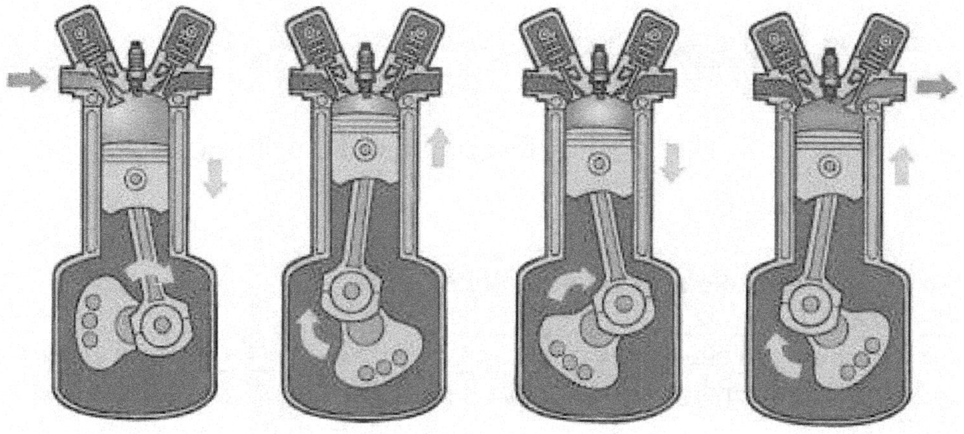

Fig.2. Cycleofoperationinfourstrokeengine

Two stroke engine:

-Nopistonstrokeforsuctionand exhaust operations

-Suctionisaccomplishedbyaircompressedincrankcaseorbya blower

-Inductionofcompressedairremovestheproductsofcombustionthroughexhaustports

-Transferportistheretosupplythefreshchargeintocombustionchamber Figure

3 represents operation of two stroke engine

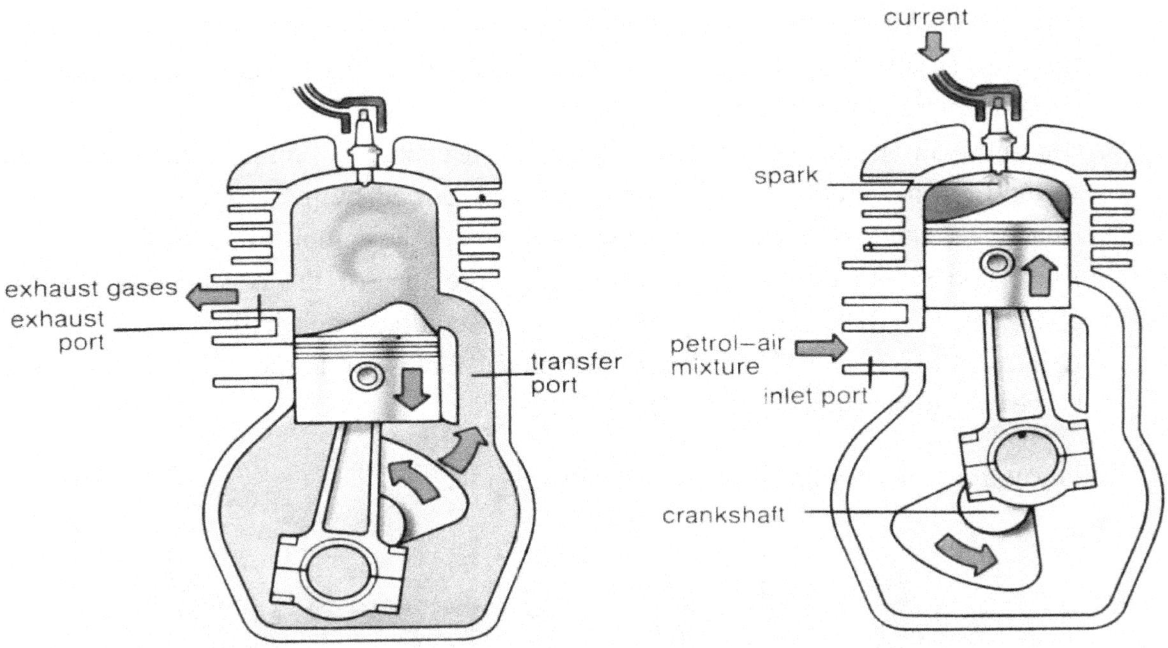

Fig.3. Cycleofoperationintwostrokeengine

Comparison of Four-stroke and two-stroke engine:

	Four-strokeengine	Two-strokeengine
f)	Fourstrokeofthepistonandtworevolution of crankshaft	Twostrokeofthepistonandone revolution of crankshaft
g)	Onepowerstrokeineverytworevolutionof crankshaft	Onepowerstrokeineachrevolutionof crankshaft
h)	Heavier flywheel due to non-uniform turning movement	Lighterflywheeldue tomore uniform turningmovement
i)	Power produce is less	Theoretically power produce istwice thanthefourstrokeengineforsamesize
j)	Heavyandbulky	Lightand compact
k)	Lessercoolingandlubrication requirements	Greater cooling and lubrication requirements
l)	Lesserrateofwearandtear	Higherrateofwearandtear
m)	Containsvalveandvalve mechanism	Containsports arrangement
n)	Higherinitialcost	Cheaperinitial cost
o)	Volumetricefficiencyismoreduetogreater time of induction	Volumetricefficiencylessduetolesser time of induction
p)	Thermalefficiencyishighandalsopartload efficiency better	Thermalefficiencyislow,partload efficiency lesser
q)	Itisusedwhereefficiencyisimportant. Ex-cars,buses,trucks,tractors,industrial engines,aeroplanes,powergenerationetc.	Itisusedwherelowcost,compactness and light weight are important. Ex-lawn mowers, scooters, motor cycles, mopeds, propulsion ship etc.

Comparison of SI and CI engie:

SIengine	CIengine
WorkingcycleisOtto cycle.	Workingcycleisdiesel cycle.
Petrolorgasolineorhighoctanefuelis used.	Dieselorhighcetanefuelisused.
Highself-ignitiontemperature.	Lowself-ignitiontemperature.
Fuelandairintroducedasagaseousmixture in the suction stroke.	Fuelisinjecteddirectlyintothecombustion chamber at high pressure at the end of compression stroke.
Carburettorusedtoprovidethemixture. Throttlecontrolsthequantityofmixture introduced.	Injector and high pressure pump used to supplyoffuel.Quantityoffuelregulatedin pump.
Useofsparkplugforignition system	Self-ignitionbythecompressionofair which increased the temperature required for combustion
Compression ratio is 6 to 10.5	Compressionratiois14to22
Higher maximum RPM due to lower weight	Lower maximum RPM
Maximumefficiencylowerduetolower compression ratio	Highermaximumefficiencyduetohigher compression ratio

Lighter | Heavierduetohigher pressures

Valvetimingdiagram:

The exact moment at which the inlet and outlet valve opens and closes with reference to the position of the piston and crank shown diagrammatically is known as valve timing diagram.It is expressed in terms of degree crank angle. The theoretical valve timing diagram is shown in Fig. 4.

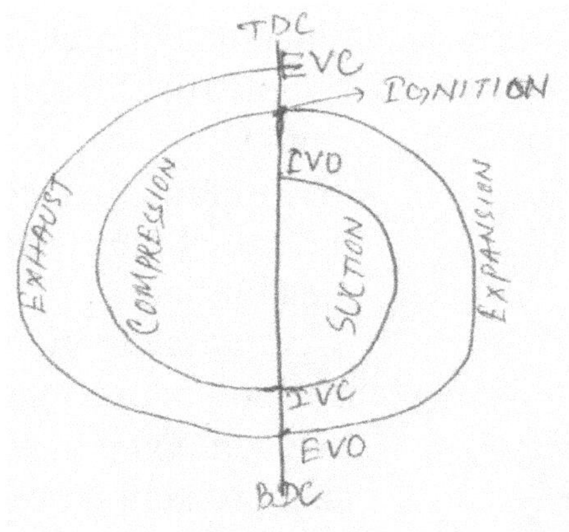

Fig.4.Theoreticalvalvetimingdiagram

Butactualvalvetimingdiagramisdifferentfromtheoreticalduetotwofactors-mechanical and dynamic factors. Figure 4 shows the actual valve timing diagram for four stroke low speed or high speed engine.

valve

-Inletvalveopens12to30°CAbeforeTDCtofacilitatesilentoperationoftheengineunder high speed. It increases the volumetric efficiency.

-Inletvalvecloses10-60°CAafterTDCduetoinertiamovementoffreshchargeinto cylinder i.e. ram effect.

Figure5representstheactualvalvetimingdiagramforlowandhighspeedengine.

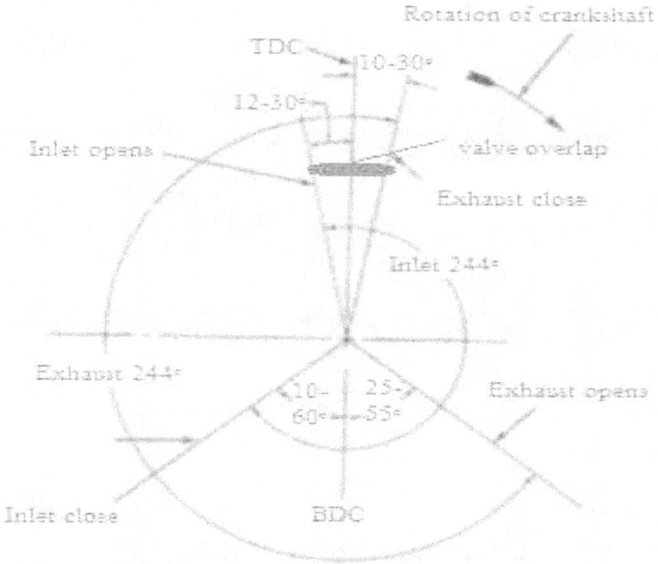

Fig.5.Actualvalvetimingdiagramforlowandhighspeedengine

Openingandclosingofexhaustvalve

Exhaust valve opens 25 to 55° CA before BDC to reduce the work required to expel out the burntgasesfromthecylinder.Attheendofexpansionstroke,thepressureinsidethechamber is high, hence work to expel out the gases increases.

Exhaustvalvecloses10to30°CAafterTDCtoavoidthecompressionofburntgasesinnext cycle. Kinetic energy of the burnt gas can assist maximum exhausting of the gas. It also increases the volumetric efficiency.

Note:Forlowandhighspeedengine,theloweranduppervaluesareused respectively

Valveoverlap

During this time both the intake and exhaust valves are open. The intake valve is opened before the exhaust gases have completely left the cylinder, and their considerable velocity assists in drawing in the fresh charge. Engine designers aim to close the exhaust valve just as the fresh charge from the intake valve reaches it, to prevent either loss of fresh charge or unscavenged exhaust gas.

Porttimingdiagram:

-Drawnfor2-strokeengine

-Novalvearrangement

-3ports-inlet,transferandexhaust

Figure6showsporttimingdiagram for2-strokeengine

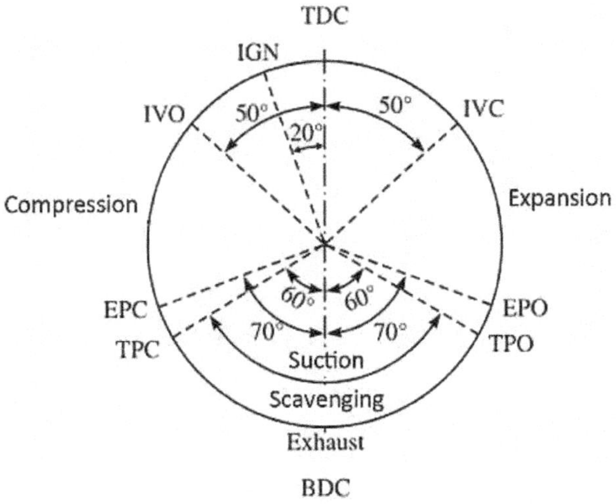

Fig.6.Porttimingdiagramfor2-stroke engine

Workingcycle:

a) ***Ottocycle-***thermodynamiccycleforSI/petrol engine
-Reversibleadiabaticcompressionandexpansionprocess
-Constantvolumeheataddition(combustion)andheatrejectionprocess(exhaust)

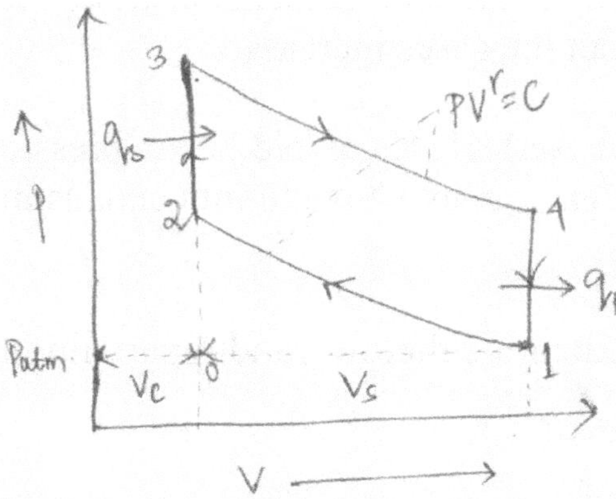

Figure 7 depicts the Otto cycle

Fig.7. Otto cycle

b) ***Dieselcycle-***thermodynamiccycleforlowspeedCI/diesel engine
-Reversibleadiabaticcompressionandexpansionprocess
-Constantpressureheataddition(combustion)andheatrejectionprocess(exhaust)
Figure 8 depicts the diesel cycle.

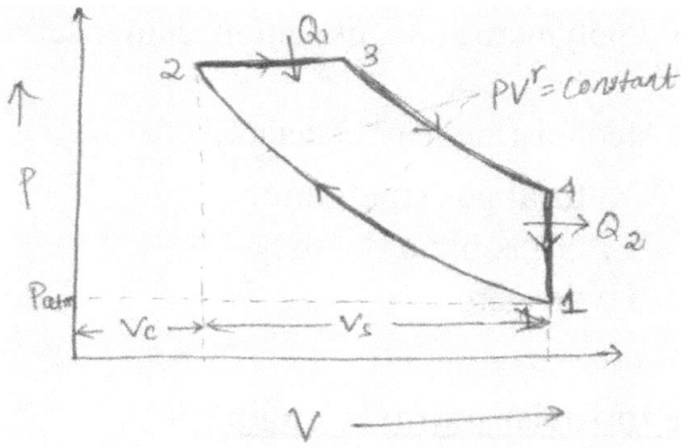

c) ***Dualcycleorlimitedpressurecycle-***thermodynamiccycleforhighspeeddiesel and hot spot ignition engine

-Reversibleadiabaticcompressionandexpansionprocess

-Constantpressureandconstantvolumeheataddition(combustion)andheat rejection process

Figure9showstheP-VdiagramofDual cycle.

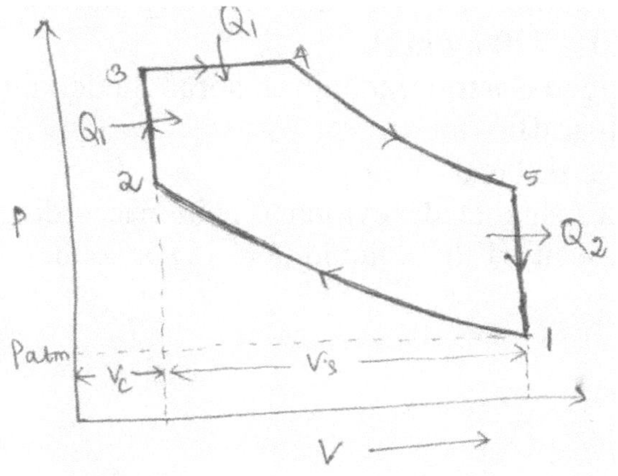

Fig.9.Dualcycle

FUELS &FUEL INJECTION

InICengines,thechemicalenergycontainedinthefuelisconvertedintomechanicalpower by burning (oxidizing) the fuel inside the combustion chamber of the engine.

FuelssuitableforfastchemicalreactionhavetobeusedinICengines,theyarefollowing types-

3. Hydrocarbonsfuelsderivedfromthecrudepetroleumbyproperrefiningprocesssuchas thermal

andcatalytic crackingmethod,polymerisation,alkylation,isomerisation,reforming and blending.

4. Alternativefuelssuchas-Alcohols(methanol,ethanol)
   Natural gas (methane)
   LPG(propane,butane)
   Hydrogen

***ClassificationofpetroleumfuelsusedforICengine:**

Liquid hydrocarbons- Engine fuels are mainly mixtures of hydrocarbons, with bonds between hydrogen and carbon atoms. During combustion these bonds are broken and new bonds are formed with oxygen atoms, accompanied by the release of chemical energy. Principal products are carbon dioxide and water vapour. Fuels also contain small amounts of S, O_2, N_2, H_2O. The different constituents of crude petroleum which are available in liquid hydrocarbons are- paraffins, naphthenes, naphthenes, olefins, aromatics.

(6) ***Paraffin-***

-Paraffinsoralkanescaningeneralberepresentedby-C_nH_{2n+2}
-Allthecarbonbondsaresinglebonds–theyare"saturated"highnumberof Hatoms,high heat content and low density (620 – 770 kg/m3)
-Thecarbonatomscanbearrangedasastraightchainorasbranchedchaincompounds.
-Straightchaingroup(normalparaffins)

- shorterthechain,strongerthebond
- notsuitableforSIengines –hightendancyforautoignitionaccordingtothevalueof "n" in the formula, theyare in gaseous (1 to 4), liquid (5 to 15) or solid (>16) state.

HexanC_6H_{14}(normalparaffin)

```
   HH     HH     H H
H–C–C–C–C–C–C–H HH
          HH     H H
```

(b) Branchedchaincompounds (isoparaffins)
WhenfourormoreCatomsareinachainmoleculeitispossibletoformisomers, theyhavethesamechemicalformulabutdifferentstructures,whichoftenleadsto very different chemical properties.

Example:Iso-octane-C_8H_{18}

(7) ***Naphthenes-***

-AlsocalledascycloparaffinsandrepresentedasC_nH_{2n}
-Saturatedhydrocarbonswhicharearrangedinacirclehavestablestructureandlow tendancy to autoignite compared to alkanes (normal paraffins)
-CanbeusedbothinSI-enginesandCI-engines
-Lowheatcontentandhighdensity(740–790kg/ m^3)

(8) ***Olefins-***

-OlefinsoralkenesarerepresentedasMonoolefins-C_nH_{2n}orDio-olefinsC_nH_{2n-2}
-OlefinshavethesameC-to-Hratioandthesamegeneralformulaasnaphthenes,their behavior and characteristics are entirely different
-Theyarestraightorbranchchaincompoundswithoneormoredoublebond.Thepositionof the double bond is indicated by the number of first C atom to which it is attached, i.e.,

CH2=CH.CH2.CH2.CH3calledpentene-1
CH3.CH=CH3 called butene-2

-Olefiniccompoundsareeasilyoxidized,havepooroxidation stability
-CanbeusedinSI-engines,obtainedbycrackingoflargemoleculeslowheatcontent and densityinthe range620–820 kg/m^3

Alkenes are such as,
Hexen(mono-olefin)

H HH H H H
H–C–C–C–C–C=C-H H HH
H

Butadien(dio-olefin)

H H H H
H– C = C–C = C–H

(9) ***Aromatics-***

-ThesearesocalledduetoaromaticsodourandrepresentedasC_nH_{2n-6}
-Theyarebasedonasix-memberedringhavingthreeconjugateddoublebonds
-Aromaticringscanbefusedtogethertogivepolynucleararomatics,PAN,alsocalled polycyclic aromatic hydrocarbons, PAH simplest member is benzene (C_6H_6)
-CanbeusedinSI-engines,toincreasetheresistancetoknocknotsuitablefor CI-enginesdue to low cetene number
-Lowheatcontentandhighdensityintherange800–850kg/m^3

***Refineryprocesses:**

Crude oil is the liquid part of the naturally occurring organic material composed mostly of HCs that is trapped geologically in underground reservoirs – it is not uniform and varies in density, chemical composition, boiling range etc. for different fields. The refinery processes involved in production of different range of fuel is shown in Fig. 10 and Fig. 11.

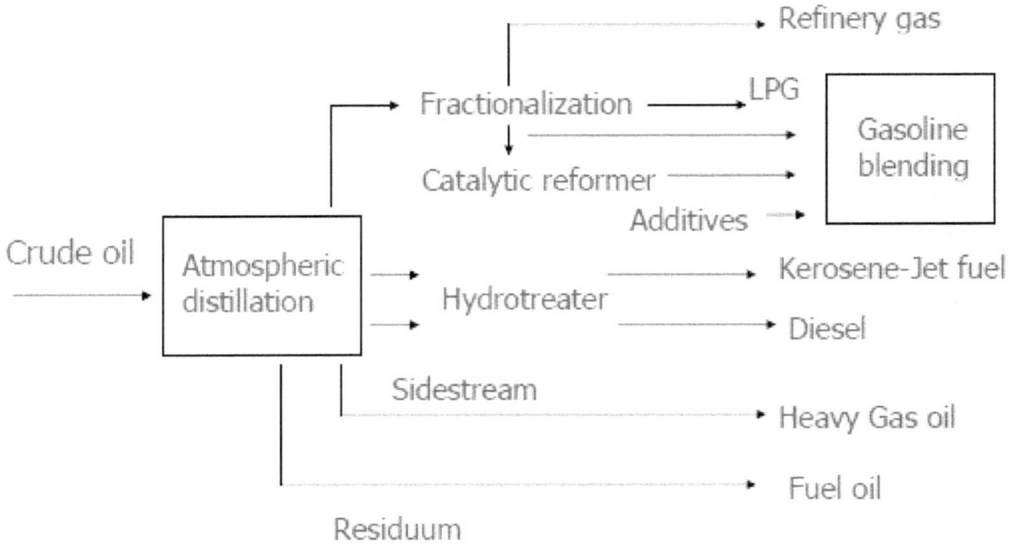

Fig.10.Refinery processes

Fig.11.Productsmadefromcrudeoil

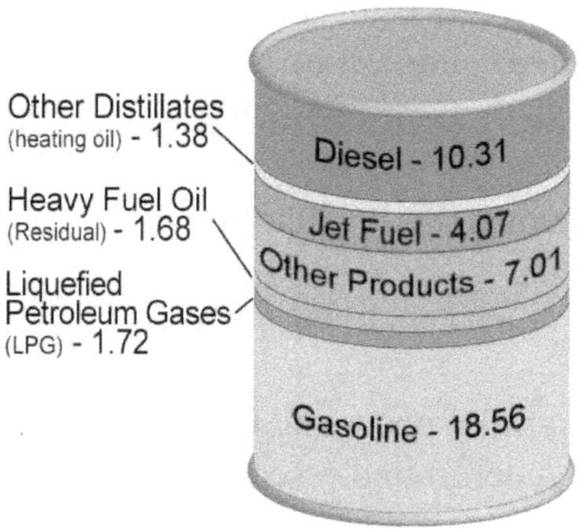

- ***Distillationprocess***

-This is the initial process used in all refineries – aims to separate the crude oil into different boiling range fractions, each of which may be a product in its own right, a blend component or feed for further processing step
-Crudeoilcontainsmanythousandsofdifferent HCs,eachhasitsownboilingpoint –lightest are gases at ambient temperature but can remain dissolved in heavier liquid HCs unless temperatureisraised,heaviestaresolidsatambienttemperaturebutstayinsolutionunless temperature is lowered.

Gasolinedistillationtemperatureis35–200°C Jet
fuel 35 - 150
Dieselfuel175 –370
Heavyfuels,oil370 –550

-Generallydistillationofcrudeoilproduces30%gasoline,20-40%dieselfuel,20%heavy fuels, 10-20 % heavy oils.

- ***Crackingprocess***

-Therearetwotypesofcrackingprocessforenginefuelproduction:thermalcracking and catalytic cracking

- **Thermal cracking:** It takes place through the creation of HC free radicals by Cto- C bond scission
-The feed is heated to around 500 - 600 °C and 70 - 100 bars and passed into a soaking chamber where cracking takes place. The cracked products are fractionated. The product is relatively unstable and requires the use of antioxidants and other treatmentstopreventgum formationinuse. Ithasrelativelypoor MON(motoroctane number).

- **Catalytic cracking:** It is the most important and widely used process for converting heavy refinery streams to lighter products – to increase the ratio of light to heavy products from crude oil.
-Compared to thermal cracking, it has higher yield, improved quality product for gasoline (not for diesel fuel) and superior economics.
-A fluidized bed of catalyst is used – feed is introduced into it. Catalyst flows from one vessel to another through a pipe (between reactor and regenerator). Cracked oil vapour pass to fractionating towers where smaller molecules are separated from heavier products (gas, catalytic naphtha, cycle oils and residue).

-Aluminiumsilicateknownaszeoliteisusedasacatalyst–hashighactivityand suppress the formation of light olefins.

- ***Alkylation:*** It is a process for producing a high-octane gasoline component (alkylate) by combining light olefins with isobutane in the presence of a strongly acidic catalyst (sulfuricor hydrofluoric acid).

- ***Isomerization:*** It is a process for converting straight chain paraffins to branch chain – used to provide isobutane feed for the alkylation process or to convert relatively low-octane quality of straight paraffins to more valuable branch chain molecules.

eg. n-pentane with RON (research octane number) 62 can be converted to isopentane with RON 92

-Process involves contacting HCs with a catalyst (platinum on a zeolite base) and separating any unchanged straight paraffins for recycle through the unit. The product is clean burning and has better RON quality.

- ***Polymerization:*** It is a process where light olefins such as propene and butenes are reacted together to give heavier olefins which have good octane quality and low vapour pressure in gasoline.

-Mostcommonlyusedcatalystisphosphoricacid
-Theproductisalmost100%olefinicandhasrelativelypoorMONcomparedwith RON.

eg. $CH_3CH_2CH{=}CH_2 + CH_3CH{=}CH_2 \rightarrow CH_3(CH_2)_4CH{=}CH_2$
butene propene heptene

*Alternative fuels:

- ***Alcohols:*** Theseincludemethanol(methylalcohol),ethanol(ethylalcohol),propanol (propyl alcohol), butanol (butyl alcohol) as compounds

-TheOHgroupwhichreplacesoneoftheHatomsinanalkane,givesthesecompoundstheir characteristic properties
-Specificheatingvalueislowerthangasoline(42 –43MJ/kg) methanol (19.7 MJ/kg) and ethanol (26.8 MJ/kg)
-Forair-fuelmixtureSHViscomperablewithgasoline(MJ/kg-mixtureatstoichiometric mixtures)
-Otheralcoholgroupssuchasdihydricandtrihydricalcoholsarenotusedasafuelin IC engines

- **Methanol**

-Canbeobtainedfromnaturalgas–hasnearandlong-termpotential
-Hashighoctanequality(130RON,95MON)
-Canbeusedinlow-concentration(5-15%)ingasolinetoincreaseoctanenumberof the mixture

Problems;

-Poorsolubilityingasoline,toxicity,lowenergycontent(abouthalfofgasoline),high latent heat of vaporization and oxygen content
-Contributetopoordriveability,incompatibilitywithsomemetals

- **Ethanol**

-Producedfrombiomass
-It is made from the sugars found in grains, such as: Corn, Sorghum, and Barley Othersourcesofsugarstoproduceethanolinclude:Potatoskins,Rice,Sugarcane, Sugarbeets,Yardclippings,Bark,Switchgrass etc.
-Hashighoctanenumber–canbeusedinlow–concentrationsingasoline
-MostoftheethanolusedintheUnitedStatestodayisdistilledfromcorn
-Scientistsareworkingoncheaperwaystomakeethanolbyusingallpartsofplants and trees rather than just the grain.
-About99%oftheethanolproducedintheUnitedStatesisusedtomake"E10"or "gasohol," a mixture of 10% ethanol and 90% gasoline.
-AnygasolinepoweredenginecanuseE10,butonlyspeciallymadevehiclescanrun on E85, a fuel that is 85% ethanol and 15% gasoline

- ***Biodiesel:***

-It is methyl or ethyl ester of a fatty acid produced from vegetable oil of edible or non edible types or animal fat or algae, by transesterification process using catalysts.
-Has better lubricating properties and much higher cetane ratings than today's low sulfur diesel fuels.
-Itsadditionreducesthefuelsystemwear.
-Can be used in the pure form (B100), or may be blended with petroleum diesel in any concentration in most diesel engines for transportation purpose.
-But, the engine may face problems, such as low temperature operation, less durability and drop in power. New diesel fuel injection systems, such as common rail systems are equipped with materials that are compatible with biodiesel (B100).
-Biodiesel offers a substantial reduction in particulate matter (25%-50%) and a marginal increase of NOx (1%-6% when it is used as an alternative fuel in a CI engine.
-The major problems associated with biodiesel are (i) poor oxidation stability, (ii) higher viscosity and density, (iii) lower calorific value, and (iv) cold flow property.
-Blends of 20% and lower biodiesel can be used in diesel engines with no, or only minor modifications.

- ***Biogas:***

-Produced by the anaerobic decomposition of organic materials such as cow dung and other waste such as cornhusks, leaves, straw, garbage, flesh of car cusses, poultry droppings, pig dung, human excreta, sewage and the plants specially grown for this purpose like water hyacinth, algae, certain types of grasses. Also any cellulosic organic material of animal or plant origin whichis easilybio-degradable isa potential rawmaterial for biogas production. - Also produced by pyrolysis and hydrogasification methods
-Contains a mixture of methane (50-60% vol), CO_2 (30-45%), hydrogen (5-10%), nitrogen (0.5-7%) and small traces of other gases such as hydrogen sulphide and oxygen
-It is a clean, but slow burning gas and having value between 5000 to 5500 kcal/kg or

38131 kJ/m^3
-Theoctaneratingofbiogasis130andignitiontemperatureis650°C
-Can be used to operate both compression ignition (diesel) and spark ignition (petrol)engines. CI engines can operate on dual-fuel (biogas+diesel) operation and pilot injection operation in which small quantity of diesel is required for igniting the mixture of air and biogas
-80%savingofdieseloilcanbeachieved
-Drawback of biogas is present of CO_2. The engine performance can be improved byreducing the CO_2 content in biogas.

- **Hydrogen:**

-Clean burning fuel and has the highest energy content per unit mass of any chemical fuels which can reduce the dependency on hydrocarbon based fuels

Production:

Most common method of producing hydrogen involves splitting water (H_2O) into its component parts of hydrogen (H_2) and oxygen (O). There are different methods to produce hydrogen-

- Steam reformation or partial oxidation of hydrocarbons such as natural gas, naphthaorcrudeoil.Itconvertsmethaneintohydrogenandcarbonmonoxidebyreactionwith steam over a nickel catalyst.
- Coal gasification- Hydrogen made from coal can probably be justified as a fuel for special applications where the unique characteristics of hydrogen can be put to advantage such as its weight or its non-polluting characteristics.
- Electrolysis- it uses electrical current to split water into hydrogen at the cathode (+) and oxygen at anode (-) [3].
- Thermo chemical method- it utilizes heat to achieve the chemical splitting of water to its elements without the need for intermediate electricity generation and without the need to use the extremity high temperature of 2500 °C or more.
- Photo-electrolysis- it uses sunlight and catalysts to split water. In this method, a current is generated by exposing on or both electrodes to sunlight. Hydrogen and oxygen gases are liberated at the 2 electrodes by the decomposition of water. A catalyst may be included to facilitate the electrode process.
- Biological and photo-biological water splitting use sunlight and biological organisms to split water.
- Thermal water splittinguses a veryhigh temperature (approximately1000 °C) to split water.
- Biomass gasification uses selected microbes to break down a variety of biomass feed stocks into hydrogen.

Utilizationofhydrogengas:

Hydrogencanbeutilizedforthefollowingpurpose:

- Residential use- hydrogen can be used in domestic cooking (stoves), radiant space heaters, electricity for lighting and for operating domestic appliances (e.g.refrigerator) which could be generated by means of fuel cells, with hydrogen gas at one electrode and air at other.

- Industrial use- hydrogen can be used as a fuel or a chemical reducing (i.e. oxygen removal) agent. It can also be used instead of coal or coal derived gases, to reduce oxide ores (iron ore) to the material (iron).

- Aircraftapplication-Theearliestapplicationofliquidhydrogen fuelisexpectedtobe in a jet air craft. Cold liquid hydrogen can be used directly or indirectly to cool the engine and the air frame surfaces of a high speed air craft.

- Electric power generation- It comprises the production of electricity by using hydrogen in fuel cell system. Hydrogen could also be used as a means for storing and distributing electrical energy. The objective of developing fuel cell power stations isto centralized and local generation of electricity.

- As an alternative transport fuel- Hydrogen is tried as an alternative fuel in internal combustionengine.Thestoichiometrichydrogenairmixtureburnsseventimesas fast as the corresponding gasoline air mixture which is a great advantage ininternal combustion engines, leading to higher engine speeds and greater thermal efficiency [2]. Hydrogen fuel used in IC engines is in automobiles, buses, trucks and farm machinery.

Methods of using Hydrogen as a fuel in CI engines

- A mixture of fuel gas and air, with an approximately constant fuel to air ratio is introduced into the cylinder intake manifold. The engine power is controlled by varying the quantity of mixture entering the cylinder by means of throttle valve. It is not safe because the mixture is formed in the manifold.

- The hydrogen is injected directly into the engine cylinder through a valve under pressure and air is inducted through another intake valve. This method is safer one, since hydrogen and air are supplied separately; an explosive mixture is occurred inside the cylinder only. The engine power output is controlled by varying the pressure of hydrogen gas from about 14 atm at low power to 70 atm at high power.

- During the intake stroke, the hydrogen gas at normal or moderate pressure is drawn through the throttle valve into the engine cylinder whereas unthrottled air is drawn in through the intake port. The variation of engine power can be achieved with adjustment of hydrogen inlet throttle. The changes in fuel proportion as well aspower is developed due to supply of un throttle air and power variation is possible because of the wide composition range over which hydrogen-air mixture can be ignited [1].

AdvantagesofusingHydrogenfuelledengine

(iii) It provides high efficiency because it utilizes a higher proportion of the energy in the fuel.

(iv) The amount of carbon monoxide and hydrocarbons in the exhaust is very small since they are originating only from the cylinder lubricating oil.

(v) Itcanbeeasilyavailablebecauseitisproducedby electrolysisofwater.

(vi) Fuelleakagetoenvironmentisnotpollutant.

DisadvantagesofusingHydrogenfuelledengine

(vi) Due to high heat release the combustion temperature may be high and also a level of nitrogen oxide is high. It can be reduced by reducing the combustion temperature by injecting water vapor into the cylinder from the exhaust.

(vii) Itrequiresheavy,bulkyfuelstoragebothinvehicleandattheservice station.

(viii) Difficultyinrefuelingandpossibilityofdetonation.

(ix) Poor engine volumetric efficiency- gaseous fuel will displace some of inlet air and poor volumetric efficiency will result.

(x) Fuelcostwouldbehighatpresentdaytechnology[2].

➢ **Natural Gas:**

-Naturalgasispresentintheearthandisoftenproducedinassociation withtheproductionof crude oil. Processing is required to separate the gas from petroleum liquids and to remove contaminants. First, the gas is separated from free liquids such as crude oil, hydrocarbon condensate, water and entrained solids. The separated gas is further processed to meet certain pipelines qualityspecifications with respect to water content, hydrocarbon dew point, heating value and hydrogen sulphide content. Generally, a gas sweetening plant removes hydrogen sulphide and other sulfur compounds

(c) Over70%ofthenaturalgasisformedbymethane.

-It is Colorless, odorless and mostly constitutes methane which is a relatively unreactive hydrocarbon.

Utilization:

-Natural gas is widely used for different purposes such as space heating, electricity generation, industrial processes, agricultural, raw material for petrochemical industry, residential, commercial and utility markets

-On a gallon equivalent basis, natural gas costs less than gasoline, diesel fuel or any other alternative fuel. Natural gas currentlysupplies over 25% of the energydemand because of its quality.

-can either be stored on board a vehicle in tanksascompressed natural gas (CNG) at pressure of 16 to 25 bar or cryogenically cooled to a liquid state (-127 °C) as liquefied natural gas (LNG) at pressure of 70 to 120 bar. As a fuel and with a single throttle body injector it works best in an engine system. LNG is used in heavy duty vehicles where use of CNG would still entail space and load carrying capacity penalties. The fuel storage system of natural gas as LNG instead of CNG is less than half the weight and volume of CNG system. So, it can be easily transportable than CNG.

➢ ***CNG(CompressedNaturalGas)***

-Naturalgasconsistsofelementsofcompressor,somesortofcompressedgasstorageand dispensing unit of CNG into vehicles

-Two types of CNG refueling system- slow fill and fast fill. In slow fill system, several vehicles are connected to the output of the compressor at one time. These vehicles are then refilled over several hours of compressor operation. In fast fill systems, enough CNG isstored so that several vehicles can be refueled one after the other, just like refueling from a single gasoline dispenser

-The storage system of CNG is arranged as several tanks in cascade form. The CNG pressure in cascade is higher than the maximum storage pressure of the cylinder on the vehicle. The cascade attempts to deliver as much of its CNG to vehicles as possible before the pressure difference decreases to where the flow rate slows dramatically. A dryer should include in most CNG refueling systems to remove water vapor, impurities and hydrogen sulphide from natural gas before it is compressed. If water vapor is present then it can condense in the vehiclefuelsystem,causingcorrosionespeciallyifhydrogensulphideispresent.CNGdriven vehicles with catalytic converter have less CO and HC emission but NOx emission is high

- ***LPG(LiquefiedPetroleum Gas)***

-LPG is available in the market in two forms- one is propane and the other is butane. Propane is popular alternative fuel because of its infrastructure of pipe lines, processing facilities and storageforitsefficientdistributionandalsoitproducesfeweremissions.Propaneisproduced as a byproduct of natural gas processing and crude oil refining

-Natural gas contains LPG, water vapor and other impurities and about 55% of the LPG is compressed from natural gas purification. LPG is a simple mixture of hydrocarbon mainly propane/propylene (C_3S) and butane/ butylenes (C_4S)

-Propaneisanodorless,nonpoisonousgaswhichhaslowestflammabilityrange.

UtilizationofLPG

LPG is used as a fuel in heating appliances and vehicles. It is increasingly used as an aerosol propellant and a refrigerant, replacing chlorofluorocarbons in an effort to reduce damage to the ozone layer

-In Europe, LPG is used as an alternative to electricity and heating oil (kerosene).It can also be used as power source for combined heat and power technologies (CHP). CHP is the process of generating both electrical power and useful heat from a single fuel source. This technology has allowed LPG to be used not just as fuel for heating and cooking, but also for de-centralized generation of electricity

-LPGhashigherpotentialasanalternatefuelforICengine.

AdvantagesofLPG

-EmissionismuchreducedbytheuseofLPG.

-LPGmixeswithairatalltemperatures.

-Uniformmixturecanbesuppliedtoallcylindersofmulti-cylinderengine.

-Enginewithhighcompressionratio(10:1)canusepropane.

-Thereiscostsavingofabout50%andlongerlifewithLPGrunningengine.

Disadvantagesof LPG

- AgoodcoolingsystemisnecessarybecauseLPGvaporizerusesenginecoolanttoprovide the heat to convert the liquid LPG to gas [2].
- Theweightofvehicleisincreasedduetotheuseofheavypressurecylinderforstoring LPG.
- AspecialfuelfeedsystemisrequiredforLPG.
- Requirementofsafetydevicetopreventaccidentduetoexplosionofgascylindersor leakage in the gas pipes.

➢ **Producer Gas:**

-Producer gas is a product of oxidation-reduction reactions of air with biomass. Biomass is chemically composed of elements C, H, O and some N and hence the oxidation results in productsofcombustionlikeCO_2andH_2O. ThemoleculesofO_2intheairoxidisesCandHto producethese products.The gases which areat high temperature dueto partial oxidation pass through a bed of charcoal (which is produced because of oxidation reaction itself) and the reduction reaction of these gases with carbon leads to carbon monoxide and hydrogen

-Volumetric composition of producer gas is CO (16-20%), H_2 (16-18%), CO_2 (8-10%) and some traces of higher hydrocarbons. Producer gas has a high percentage of N_2, since air is used. So it has a low heat value. Density of producer gas is 0.9 to 1.2 kg/m^3

-Producer gas is used in reciprocating engines and furnace. It needs little air to burn stoichiometrically and raise the temperature to a value of 1500 K at normal temperature and pressure. It is also used to power gas turbines (which are well-suited to fuels of low calorific value), spark ignited engines (where 100% petrol fuel replacement is possible) or diesel internal combustion engines (where 40% - 15% of the original diesel fuel) is still used to ignite the gas.

➢ **<u>BlastFurnaceGas:</u>**

It is a byproduct of melting iron ore in steel plants. It principally consists of CO and contains low heat value similar to producer gas. It consists of about 60% nitrogen, 18-20% CO_2 and some amount of oxygen which are not flammable. It may be combined with natural gas or coke oven gas before combustion or a flame support with richer gas or oil is provided to sustain combustion. The auto ignition temperature of blast furnace gas is approximate 630 °C and it has Lower Explosive Limit (LEL) of 27% &Upper Explosive Limit (UEL) of 75% in an air-gas mixture at normal temperature and pressure. The gas is hazardous due to higher concentration of carbon monoxide [50].It should be cleaned properly because it contains lotof dust particles. Blast furnace gas depends upon types of fuel used and method of operating the blast furnace.

- **CokeOven Gas:**

It is produced during the making of coke. It is also resulting from oxidation-reduction reactions of coal or coke with air and sometimes steams.It depends upon the type of coal used and operation method of oven. The composition of coke oven gas is H_2 (54% vol), CH_4 (24%), CO (8%), CO_2 (6%) and some traces of higher hydrocarbon and nitrogen. With the application of heat the heavier hydrocarbons are cracked and volatile portionof coal is driven offand results in high composition of$H_2$and CH_4. Its heat value per cubic meteris onlyabout one half that of natural gas and density is 0.40 kg/m^3.

***GeneralFuel Specifications:**

Differentpropertiesoffuels have,

Relativedensity(specificgravity) Fuel
composition
Specificheatingvalue Flash
point
Viscosity
Surfacetension
Freezing point

iii) **Relativedensity(specificgravity):**
Itisrelatedtothemeasurementoftheratiooftheweightofa givenvolumeoffueltothe weight of the same volume of water, both at 20 ^{O}C and 101.325 kPa

Forgasoline,therelativedensityis around0.72to0.78-whichisequivalenttoanAPI (American petroleum institute) range of 65 to 50,

ρ=700−800[kg/m^3],forunleadedgasolinethisvalueishigherduetothearomatics For

diesel fuel, ρ = 830 − 950 [kg/m^3]

iv) **Fuel composition:**

C and H: carbon content of aromatics is around 89 %, and of paraffins and naphthenes is around 86 %
Benzene: maxallowableconcentration isspecifiedbecauseitishighlytoxicmaterial, the level is 5 %
Sulphurcontent: HC fuelscontainfreesulphur,hydrogensulphideandothersulphur compounds which are objectionable it is a corrosive element that can corrode fuel lines, carburettor and injection pump. It will unite with oxygen to form sulphur

dioxide, which in presence of water at low T, forms sulphurous acid.

-It has low ignition T, promote knock in SI engines. limited to approx 250 ppm (50 ppm is aimed for low pollutant emitting vehicles)
Gum deposits: gasoline with unsaturated HCs forms gum in the engine, paraffin, naphthene and aromatic HCs also form some gum – it causes operating difficulties, sticking valves and piston rings, deposits in the manifold etc.
Water: both dissolved and free water can be present in gasoline, free water is undesirable because it can freeze and cause problems. Dissolved water is usually unavoidable during manufacture.
Lead: for leaded and unleaded gasoline max lead content is specified, lead causes pollution and destroys catalytic converters in the exhaust system.
Manganese: used for antiknock in gasoline (MMT), max amount is specified, 0.00025 to 0.03 gMn/L
Oxygenates:oxygenatedcompoundssuchasalcoholsareusedingasolinetoimprove octane rating.
In USA gasohol (10% ethanol contains 3.5% oxygen), TBA and methanol up to 3.5% oxygen methanol up to 5% volume, MTBE up to 15% are used. In EC monoalcohols and ethers withatmboilingpoints lower thanthefinal atm boilingpointofgasolinein thestandardscanbeused.Higherconcentrationsrequiremodificationsonthevehicles - carburator or fuel injection system must be modified to compensate for the oxygen content of the fuel. Blends with 15% methanol can be used.

(C)Specificheatingvalue:

-Specificheatingvalue,$H_u$isameasureoftheenergycontentofthefuelperunitmass(kJ/kg or kcal/kg)

-Gaseousfuelsspheatingvalueisgivenintermsofenergycontentperunitvolume(kJ/liter or kJ/m^3, kcal/m^3)

-InICengineslowerheatingvalueisgivenasthecombustionproductscontainwaterin vapour form.
Forgasolineanddieselfuel
H_u=42000-44000kJ/kgorH_u=10200-10500 kcal/kg
-Heatingvalueofthecombustibleair-fuelmixtureisadecisivefactorforengine performance.

+ **Flashpoint:**

-Flashpointisthelowesttemperatureofasampleatwhichthefuel vapourstartstoignite when in contact with a flame (ignition source).
-Marcussonmethod–fuelcontainerisslowlyheated,whilethefuelvapourisincontactwith an open flame – T is measured
-Forgasolineitis25°C,dieselfuel35°Candheavydiesel65°C

+ **Viscosity:**

-Viscosity is an important parameter for CI engines, also influences fuel metering orifices since Re is an inverse function of fuel viscosity lower the viscosity, smaller the diameter

of the droplets in the spray.
-Below certain limits, low viscosity increases the leaks in the fuel system. It is a strong function of T – must be given at certain T values
at50°C,1.5–5.0 Engleror0.5 to0.6 centistokes

+ **Surfacetension:**

-Surfacetensionisaparameterwhicheffectstheformationoffueldropletsinsprays
-increasingthesurfacetensionwillreducemassflowandair-fuelratioingasolineengines
-lowerthevalue,smallerthedropletdiameter
dieselfuelvalueisintherangeof0.023–0.032N/m and for
gasoline it is 0.019 – 0.023 N/m

+ **Freezingpoint:**

-theprecipitationofparaffincrystalsinwintercanleadtocloggedfilters.Itcanbeprevented by either removing paraffins from the fuel or adding flow improvers (additives).
-Antifreezingpropertiesaredeterminedbyits filterability.
-Forgasolinefreezingpointis–65°Candfordieselfuel –10°C

***Importantfuelspecificationsforgasoline(a)**

Gasoline volatility:

Benzeneforexamplehasvaporpressureof0.022MPaat38OCinaclosedcontainerof38 OC, benzene evaporates until the partial p
hasavalueof0.022MPa,IfTisraisedto80.5OC,thensaturationp will be
0.1 MPa and will be constant during the boiling
ForgasolineitisnotpossibletoindicateasinglevalueofevaporationTorvaporpressure. Gasoline contains large number of compounds - up to about 400

Ithasasmoothdistillationcurve-withgoodfractionation efficiency

Lowfractionationefficiencyeffectsengineperformanceatdifferentoperatingconditions:
If distillation curve is displaced downward, gasoline becomes more volatile - poor hot start, vapor lock, high evaporative losses

Gasolinedistillationcurve:
Gasolinehavingboilingpointupto70OC controlseaseofstartingandhotweatherproblems such as vapor lock
Mid-range controls the driving in cold weathers, particularly at warm up period of engine. It also influences the ice forming in carburetor.
Back end of the curve contains all the heavier, high boiling point compounds and these have high heat content - they are important in improving fuel economy for fully warmed upengine.
Some of the heavier compounds may pass into the crankcase and dilute the crankcase oil. They are not readily combusted as the lighter compounds - cause combustion chamber deposits.

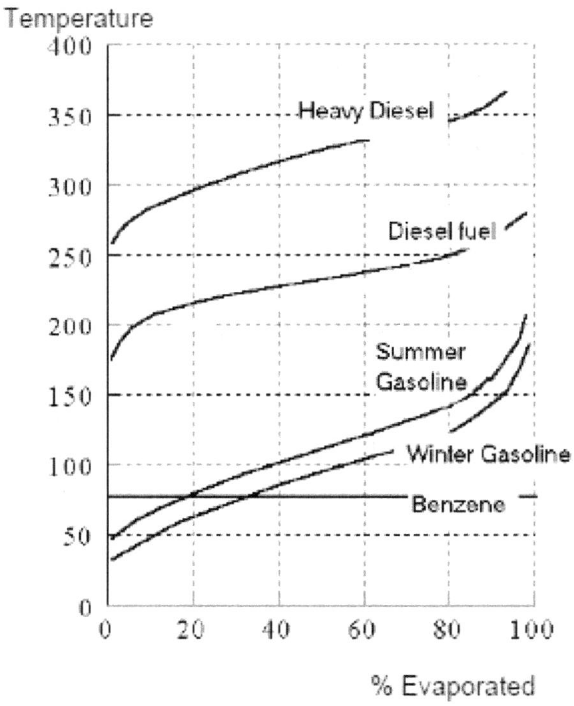

Fig.12.Distillationcurve

%10evaporationpointshouldbeatlowTforstartupatcoldtemperatures -athotweathers this may cause problems - vapor lock.
50%evaporationshouldbe slightlyabove100OCatsummerandslightlybelow100OCat winter. For warmed up engine conditions this point is not important.
90%evaporationmustnotbehigh-producesfuelfilmonintakemanifoldwallsanddilutes lubricating oil. Back end of the curve must not exceed 215 OC.
Gasolinevolatilityshouldbearrangedaccordingtoweatherconditions-particularlyambient T.Altitudehassomeminoreffectduetopressurechanges.
Itisalsoeffectedbythecharacteristicsofthevehicleitself(drivability,fuelsystemdesign etc).

Cold starting:

ForSIenginestostart,A/Fratiomustbewithintheignitablerange,ieingeneralmustbe between 7:1 to 20:1 by weight.

When the engine is cold, it is difficult to ignite lean mixtures, because fuel may not vaporize sufficiently - under these conditions the mixture is rich to bring it to ignitable range. This is done by the injection time or by the use of a choke with carburetted engines.

Measurementofgasolinevolatility:

TestsusuallydefineReidVapourPressure -ASTMDistillationtestandVapour/Liquidratio.

Reidvapourpressure-obtainedatair-to-liquidratioof4:1andtemperature37.8°C.

Fuel isfilled intoametalchamber whichisconnected toan airchamber and that isconnected to a pressure gauge. The apparatus is immersed in water bath at 37.8 °C and is shaken until constant p is obtained - Reid VP

ForgasolineallowableRVPis0.7barinsummerand0.9barinwinter(at37.8°C)

ASTM Distillation procedure - distillation rate is controlled by the heat input - distillation curve is plotted (temperature vs % evaporated).

(b)Antiknockqualityofgasoline:

-Knock occurs when the unburnt gases ahead of flame front (the end gases) spontaneously ignite causing a sudden rise in pressure accompanied by a characteristic pinging sound – this result in a loss of power and can lead to damage the engine.

-Combustion chamber shape, spark plug location, ignition timing, end gas temperatures, in cylinder gas motion, air-fuel ratio of the mixture, fuel specifications etc. effects the occurrence of knock.

-Compression ratio of the engine also strongly effects knock. The higher the CR, the better the thermal efficiency - but the greater the tendency for knock to occur.

-Criticalcompressionratio-whenknockstarts.Sohigherfueloctanequalityisrequired.

-Autoignition of the end gases causes a rapid increase of p, producing p waves whichresonate in the combustion chamber at a frequencyofbetween 5000 - 8000 Hz, dependingon the geometry of the chamber

Knock results in an increase of T in the cylinder and causes a severe damage to engine components like cylinder head gasket, piston, spark plugs etc.

Octanenumber:

-In 1929 Octane scale was proposed by Graham Edgar. In this scale two paraffinic HCs have been selected as standards (PRF, primary reference fuels)- iso-octane (2-2-4 trimethyl pentane) with veryhigh resistance to knock (arbitrary assigned a value of 100) and n-heptane with extremely low knock resistance (assigned a value of 0).

-Octane number of the fuel is the volume percentage of iso-octane in a blend with n-heptane (PRF), that shows the same antiknock performance as test fuel tested in standard engine and standard conditions.

-Test engine for determining Octane values, was developed by Cooperative Fuel Research Committee (CFR). It is a single cylinder, variable CR engine.

-Two different test conditions specifies the Research Octane Number (RON) and the Motor Octane Number (MON)

-AntiknockIndex=(RON+MON)/2

-TEL is added to the PRF to increase the ON above 100 or n-heptane is added to the sampleto reduce ON below 100, then nonlinear extrapolation is applied
ON can be increased by antiknock agents - at less expense than modifying HC compositionby refinery process.
Mosteffectiveagentsareleadalkyls- TEL
- tetraethyl lead, $(C_2H_5)_4$ Pb TML -
tetramethyl lead,

MMT
additionofabout0.8gleadperlitre,providesagainofabout10ONingasoline

***Importantfuelspecificationsfordiesel(a)**

Viscosity:

-Viscosityofafluidindicatesitsresistancetoflow -highertheviscosity,thegreaterthe resistance to flow.
-Itmaybeexpressedasabsoluteviscosity(Poise,P)orkinematicviscosity(stoke, St).
-Itvariesinverselywithtemperature,usuallygivenat20 -40°C
-Fuelatomizationdependsonviscosity 2 -
8 mm^2/s (cSt) at 20 ^{O}C

-Lowertheviscosity,smallerthediameterofthedropletsinthespray

- **Surfacetension:**

-Surfacetensionisaparameterwhicheffectstheformationoffueldropletsinsprays
-Lowerthevalue,smallerthedropletdiameter

-Dieselfuel0.023–0.032 N/m

- **Cetanenumber:**

-Cetanenumberisusedtospecifytheignitionqualityofdieselfuel.
-Running on low Cetane number will produce cold start problems. Peak cylinder pressure, combustion noise and HC emissions will increase -more fuel will be injected before ignition, less time for combustion.
-HigherCNresultsina soonerignition -extremelyhighCNmayignitebeforeadequateFuel- Air mixing can take place - higher emissions. Power output can be reduced if burning starts too early.

Measurementofcetanenumber:
-Cetanenumberis measured bycomparingthe “ignition delaytime” ofthe sample fuel witha mixture of cetane ($C_{16}H_{34}$) and alphamethyl naptane ($C_{10}H_7CH_3$). The Cetane percentage in the mixture gives the CN of the sample fuel.
-CNofthereferencefuelcetaneisarbitrarylysetat100,andofalphamethylnaptaneat0.
-CFR engine is used to measure the compression ratio at which ignition starts. CR isgraduallyincreased while the engine is driven byan electric motor - a curve of CN vs critical CR is obtained.
-Inletairtempis30 ^{o}Candcoolingwatertempisat100^{o}C

AneasierandpracticalmethodtoobtainCetaneNumberisbycalculatingtheDieselIndex. Increasing the DI, increases the tendency to ignite.
-APisobtainedbyheatingequalamountsofannilinanddieselfuel.Whilecoolingdown,the temp at which the annilin separates from the mixture is the AP

-Cetanenumberisintherangeof,
50 - 60 for high speed Diesel engines
25 - 45 for low speed Diesel engines
Normal Diesel fuel CN is 40 - 55DIof50givesaCNofaround50

***Carburetion:**
Theprocessofpreparingacombustiblefuel-airmixtureoutsideenginecylinderinSIengine is known as carburetion.
Importantfactorswhichaffecttheprocessofcarburetionaregivenbelow;
-timeavailableforthemixturepreparationi.e.atomisation,mixingandthe vaporisation
-Temperatureoftheincomingair
-qualityofthefuelsupply
-designofcombustionchamberandinductionsystem

***Mixturerequirementsforsteadystateoperation:**

Threemainareasofsteadystateoperationofautomotiveenginewhich requiredifferentair fuel

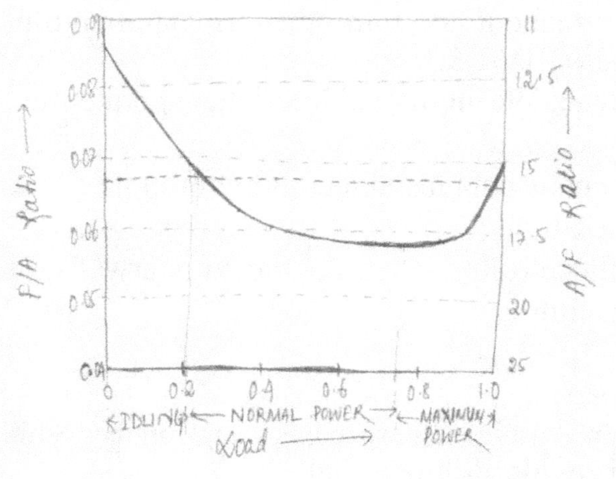

ratio are discussed below,

Fig.13.Mainareasofautomotiveengineoperation

4) ***Idlingandlow load:***
-fromnoloadtoabout20%ofratedpower
-Noloadrunningmodeiscalledidlingcondition
-verylowsuctionpressuregiverisetobackflowofexhaustgasesandairleakage
-increasestheamountofresidualgasesandhenceincreasethedilutioneffects

-Richmixturei.e.F/Aratio0.08orA/Fratio12.5:1providesmoothoperationoftheengine

5) ***Normalpowerrangeorcruisingrange:***
-fromabout20%to75%ofratedpower
-dilutionbyresidualgasesaswellasleakagedecreases,hencefueleconomyisimportant consideration in this case
-maximumfueleconomyoccursatA/Fratioof17:1to16.7:1
-mixtureratiosforbesteconomyareverynear tothemixtureratiosforminimum emissions

6) ***Maximumpowerrange:***
-fromabout75%to100%ofratedpower
-mixturerequirementsforthemaximumpowerisarichmixture,ofA/Fabout14:1orF/A 0.07
-Richmixturealsopreventsthe overheatingofexhaustvalveathighloadandinhibits detonation
-inmulti-cylinderenginetheA/Fratioareslightlylower

*Mixturerequirementsfortransientoperation:

-Carburettorhastoprovidemixturefortransientconditionsunderwhichspeed,load, temperature, or pressure change rapidly
-evaporationoffuelmaybeincompleteinthetransientcondition,quantityoffuelmaybe increasing and decreasing

(a) Startingandwarmuprequirements:
-enginespeedandtemperaturearelowduringthestarting oftheenginefrom cold
-duringstartingveryrichmixtureabout5to10timesthenormalamountofpetrolis sulpplied i.e.F/Aratio0.3to0.7orA/Fratio3:1to1.5:1
-mixtureratioisprogressivelymadeleanertoavoidtoorichevaporatedfuel-airratioduring warm up condition
-toohighvolatilitymayformvapourbubblesinthecarburettorandfuellinesparticularly when engine temperatures are high
-toolowvolatilitymaycausethepetroltocondenseonthecylinderwalls,dilutingand removing the lubricating oil film

(b) Accelerationrequirements:
(d) Accelerationrefertoanincreaseinenginespeedresultingfromtheopeningofthe throttle
-accelerationpumpisusedtoprovideadditionalfuel

*SimpleCarburettor:
-provideair-fuelmixtureforalloperatingconditions
-Carburettordepressionispressuredifferentialinthefloatchamberandventurethroatwhich causes discharge of fuel into the air stream
-flowiscontrolledbysmallholeoffuel passage
-pressureatthethroatatthefullyopenthrottleconditionliesbetween4and5cmofHgand seldom exceeds 8 cm Hg

-petrolengineisquantitygoverned
-Drawbackofsimplecarburettoristhatitprovidestoorichandtooleanmixturedueto vacuum

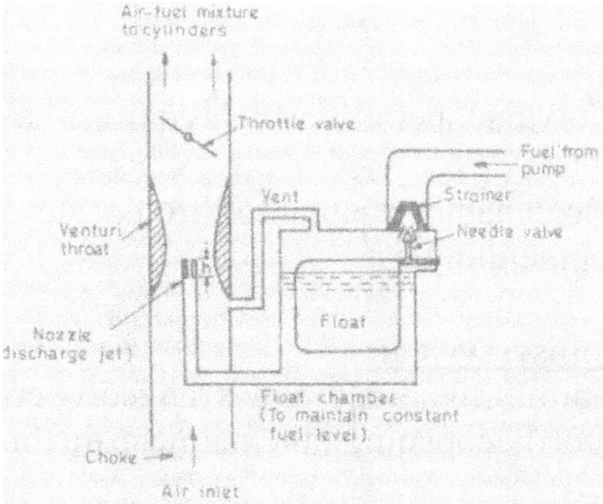

created at the throat is too high and too small which is undesirable

Fig.14.Asimple carburettor

***CompleteCarburettor:**

Additionalsystemsusedwithsimplecarburettorcanhelptheenginetooperateatall conditions, which are given below,

(i) Mainmeteringsystem:

-provideconstantfuel-airratioatwiderangeofspeedsandloads.
-mainlybaseduponthebesteconomyatfullthrottle(A/Fratioabout15.6:1) The different metering systems are,

Compensatingjetdevice:-Additiontothemainjet,acompensatingjetisprovidedto provide the leanness effect

Emulsiontubeorairbleedingdevice:

-mixturecorrectionisdonebyairbleedingalone
-inthisarrangementmainmeteringjetisfittedabout25mmbelowthepetrollevelwhichis called as submerged jet

Backsuctioncontrolorpressurereductionmethod:

-in this arrangement large vent line connects the carburettor entrance with the top of the float chamber and another small orifice line is connected with the top of the float chambers with venture throat
-itcreatespressuredifferencesaccordingtoengineoperatingconditions

Auxiliaryvalvecarburettor:

-Valvespringofauxiliaryvalveliftthevalveduringincreaseofengineloadwhichincreases the vacuum at venture
-Allowsmoreadmittancemoreadditionalairandthemixtureisnotoverrich

Auxiliaryportcarburettor:
-openingofbutterflyallowsmoreairinductancewhichdecreasesquantityoffueldrawn
-usedinaircraftcarburettors

(ii) Idlingsystem:
-Idlingjetisaddedfortheidlingandlowloadoperationwhichrequiresrich mixturesof about A/F ratio 12:1
-consistsofsmallfuellinefromthefloatchambertoapointofthrottleside
-gradualopeningofthrottlemaystoptheidlingjet

(iii) Powerenrichmentoreconomisersystem:
-thissystemprovidestherichermixtureformaximumpowerrangeofoperation
-Ithasmeterrodeconomiseroflargeorificeopeningtothemainjetasthethrottleisopened beyond a certain point
-rodistaperedor stepped

(iv) Accelerationpumpsystem:
-Engineaccelerationconditionorrapidincreasein enginespeedmayopenthethrottlerapidly which will not able to provide rich mixture
-accelerationpumpofspringloadedplungerisusedforfuelsupply

(v) Choke:
-Richmixtureisrequiredduringcoldstartingperiod,atlowcranckingspeedandbeforethe engine warmed up condition
-butterflytypevalveorchokeisusedbetweentheentrancetothecarburettorandventure throat to meet the requirement
-springloadedby-passchokeisusedinhigherspeeds

***Carburettortypes:**

(i) Openchoke:Zenith,solexandcarter
Constantvacuumtype:S.U.carburettor

(ii) updraughttype
Horizontalordowndraught:mixtureisassistedbygravityinitsoassagetotheengine induction

(a) Solexcarburettor:
-provideeaseofstarting,goodperformance,andreliability
-usedinFiatandstandardcarsandWilleyjeep
-Bi-starterisusedforcold starting
-wellofemulsionsystemisusedforidlingandslowrunningcondition
-diaphragmtypeaccelerationpumpisusedforincreasingspeedcase

(b) Cartercarburettor:
-downdrafttypecarburettorusedinjeep
-hastripleventurediffusingtypechokeinwhichsmallestliesabovethelevelfloatchamber, other

two below the petrol level, one below other
-multipleventuriesresultinbetterformationofthemixtureatverylowspeedscausingsteady and smooth operation at very low and high engine speed
-mechanicalmeteringmethodisused
-chokevalveisprovidedintheaircircuitforcoldstarting
-plungertypeaccelerationpumpisused

(c) S.U.carburettor:
-constantair-fuelratioismaintainedduetovacuumdepression
-hasonlyone jet
-noseparateidlingjetoraccelerationpump
-constanthighvelocityairacrossthejetmayavoidtheuseofidlingjet
-jetleverarrangementprovidestherichmixtureincoldstarting
-usedinmanyBritishcarsandHindustanambassadorcar

Drawbacksofmoderncarburettor:
-impropermixtureproportioninmulti-cylinderengine
-lossofvolumetricefficiencyduetoobstructionofflowofmixturefromchoketubes,jets, throttle valve etc.
-wearofcarburettorparts
-Freezingatlowtemperature
-surgingwhencarburettoristiltedorduringacrobaticsinaircraft
-backfiringinfuelpipeline

Petrol injection:

-toavoidaboveproblemofmoderncarburettor,petrolinjectionisusedlikeindieselengine
-petrolinjectedduringthesuctionstrokeintheintakemanifoldatlowpressure
-injectiontimingisnotmuchcriticalaslikeindieselengine
-continuousinjectionandtimedinjectionmethodsareused

Continuousinjection:

-fuelissprayedatlowpressurecontinuouslyintotheairsupply
-amountoffuelisgovernedbyairthrottleopening
-insuperchargedengine,fuelinjectedintheformofmultiplesprayintothesuctionsideof the centrifugal compressor
--provideefficientatomisationoffuelanduniformmixturestrengthtoallcylinder
-highervolumetricefficiency
-onefuelinjectionpumpandoneinjector

Timedinjectionsystem:

-similartohighspeeddiesel engine
-componentsarefuelfeedorliftpump,fuelpumpanddistributorunit,fuelinjectionnozzles and mixture controls
-mixturecontrolsareautomaticforallengineoperatingconditions

(i) Multipleplungerjerkpumpsystem:

-pumpwithseparateplungerandhighinjectionnozzlepressureforeachcylinder
-100to300barpressure
-measuredquantityoffuelfordefinitetimeandoverdefiniteperiodisdelivered

(ii) Lowpressuresinglepumpanddistributorsystem:

-singleplungerorgearpumpsupplyfuelatlowpressuretoarotatingdistributor
-pressureabout3.5to7 bar

(a) Lucaspetrolinjectionsystem:

-firstlyusedinracingcar
-singledistributorsystemwithnovelmeteringdevice
-linepressureismaintainedat7bar
-meteringdistributorandcontrolunitdistributestherequiredamountoffuelatcorrecttime and interval
-hasshuttlearrangementsformeteringunit
-inaircraftenginetwoinjectorsandsparkplugprovidedfordirectinjectionoffuelin

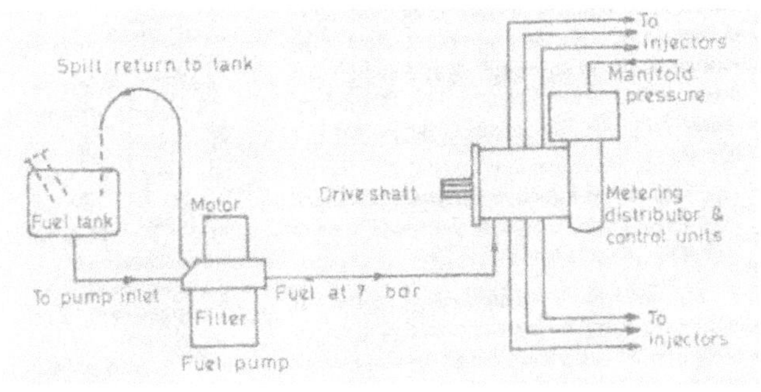

combustion chamber

Fig.15.Lucaspetrolinjectionsystemfor6-cylinderpetrolengine

(b) Electronicfuelinjection

Fueldeliverysystem:

-electricallydrivenfuelpumpdrawsfuelfromtankstodistribute
-fuelandmanifoldpressurekeptconstantbypressureregulator

Airinductionsystem:

-airflowmetergeneratevoltagesignalaccordingtoair flow
-coldstartmagneticinjectionvalvegivegoodfuelatomisationandalsoprovideextra fuel

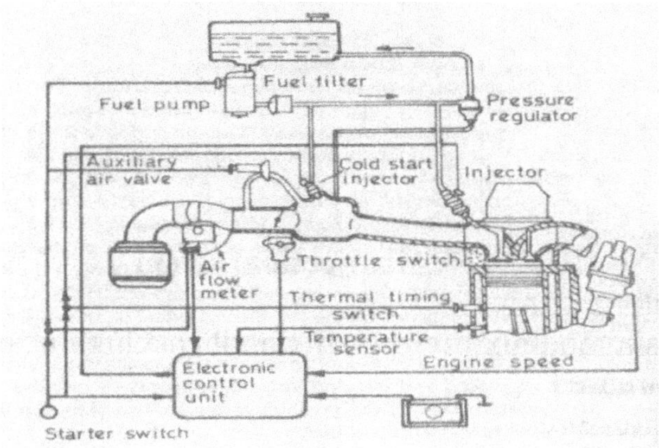

during warm up condition

Fig.16.Electronicfuelinjectionsystem-L-Jetronicwithairflowmeter

Electroniccontrolunit (ECU):

-sensorsformanifoldpressure,enginespeedandtemperatureatintakemanifold
-sensormeasuresoperatingdatafromlocationsandtransmittedelectricallytoECU

Injectiontiming:

-injectedtwiceforeveryrevolutionofcrankshaft
-triggeringofinjectors

***Dieselinjectionsystem:**

Requirementsofdieselinjectionsystem:

-fuelmustintroducepreciselydefinedperiodofcycle
-amountsmeteredveryaccurately
-rateofinjectionmeetdesiredheatreleasepattern
-quantitiesoffuelmeetchangingspeedandloadcondition
-goodatomisationoffuel
-goodspraypatternforrapidmixingoffuelandair
-nodribblingandafterinjectionoffueli.e.sharpinjection
-injectiontimingsuitsthespeedandloadrequirements
-distributionoffuelinmulti-cylindershould uniform
-weight,sizeandcostoffuelinjectionsystemshouldbeless

Typesofdieselinjectionsystem:

(a) Airinjectionsystem:

-fuelsuppliedthroughcamshaftdrivenfuel pump
-fuelvalveisalsoconnectedwithhighpressureairlinetoinjectintocylinder
-multi-stagecompressorwhichsupplyairatapressureofabout60to70bar

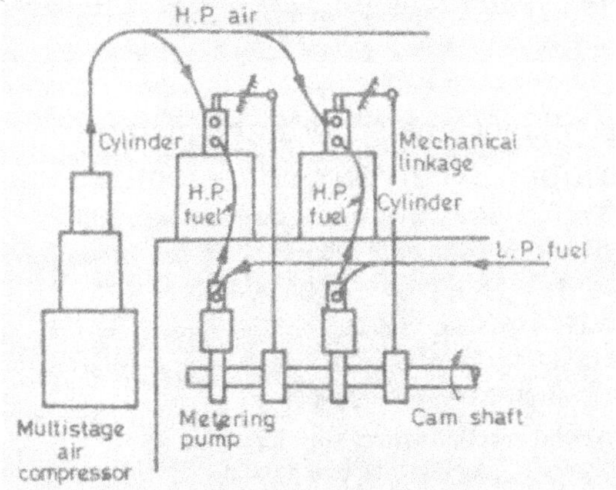

Fig.17.Airinjectionsystem

-blastairsweepsthefuelalongwithit
-goodatomisationresultsingoodmixtureformationandhencehighmeaneffectivepressure
-heavyandviscousfuelsareused
-fuelpumprequiresmallpressure
-butitiscomplicatedduetocompressorarrangementandexpensive
-bulkyengineandlowbhp
-overheatingandburningofvalveseat

(b) Solidinjectionsystem:

-Fueldirectlyinjectedtocombustionchamberwithoutprimaryatomisationtermedassolid injection.
-Alsoknownasairlessmechanicalinjection
-2units-pressuriseandatomisingunit
3differenttypeswhicharedescribedbelow,

(i) Individualpumpandinjectororjerkpumpsystem:

-separatemeteringandcompressionpumpisusedforeachcylinder
-reciprocatingfuelpumpisusedtometerandsettheinjectionpressureofthe fuel
-heavygeararrangementswhichgivesjerkingnoise,hencenameisgivenisjerk pump
-jerkpumpisusedformediumandhighspeeddieselengines

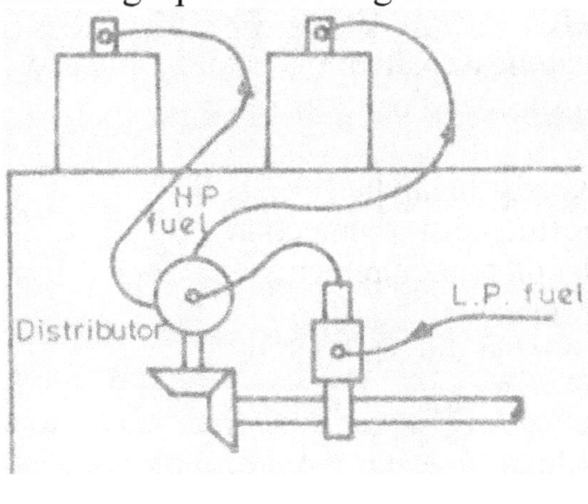

Fig.18.Individualpumpandinjectororjerkpumpsystem

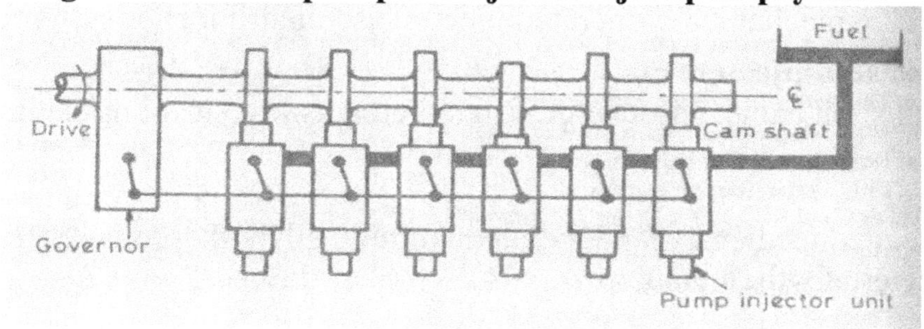

Fig.19.Unit injector

(ii) Commonrailsystem:
-highpressurefuelpumpdeliversfueltoanaccumulatorwhosepressureisconstant
-plungertypeofpumpisused
-drivingmechanismisnotstressedwithhighpressurehencenoiseisreduced
-commonrailorpipeisconnectedinbetweenaccumulatoranddistributingelements
-separatemeteringandtimingelementsconnectedtoautomaticinjector
-self-governingtype

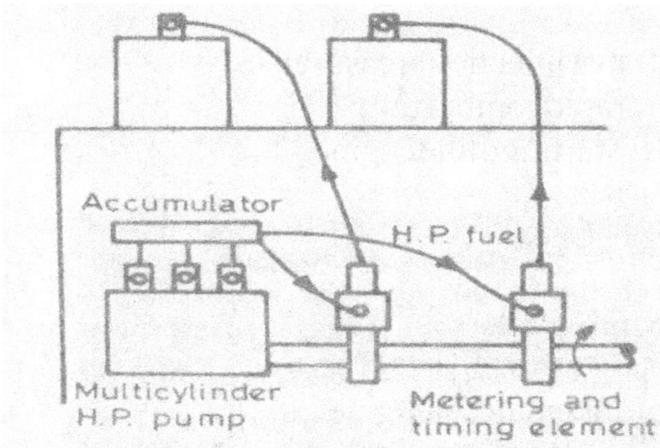

Fig.20.Commonrailsystem

(iii) Distributorsystem:
-fuelpumppressurises,metersandtimesthefuelsupplytorotatingdistributor
-numberofinjectionstrokespercycleforthepumpequalstothenumberofcylinder
-Onemeteringelementwhichensureuniformdistribution

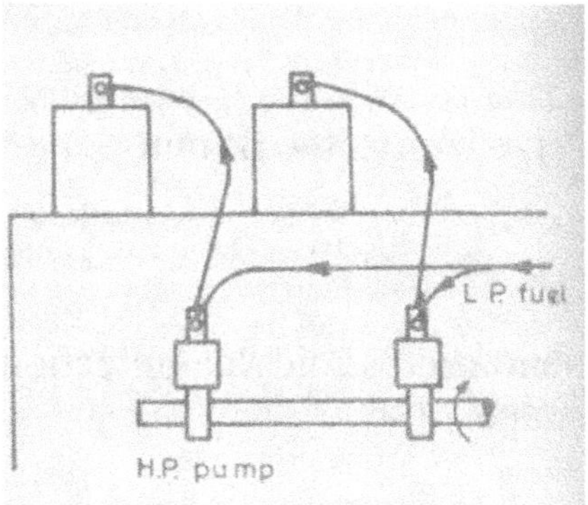

Fig.21.Distributorsystem

➢ **Fuel Injectors**
3maintypesoffuelinjectors,

Blastinjector:
-thesearesupersededbymechanicallyoperatedinjectorsusedinairinjectionsystem

Mechanicallyoperatedinjector:
-consistofasetofcamshaft,camsandrockergearandothercamsforcontrollingthetiming of the fuel injection

Automaticinjector:
-consistsofspringloadedneedlevalveandoperatedhydraulicallybythepressureoffuel
-quanityoffuelismeteredbythefuelpump

Typesof nozzles:
(a) Dependsonthetypeofcombustion chamber,

Opencombustionchamber:
-fuelseeksair
-airswirliscreatedduetoinclinedinductionport
-multi-holenozzleinjectsfuelatapressureofabout200to300bartoslowmovingair
-providegoodcoldstartingperformanceandimprovedthermalefficiency

Pre-combustionchamber:
-airvelocityisverymuchhigh
-singleholenozzlewith65to100barinjectionpressureisused
-usedinhighspeedengineduetorapidcombustion
-externalheatingdeviceforeasystartingoftheengine

(b) Openandclosedtypeofnozzle, Open type:
-consistsoffuelorificesandopento burner
-cheapandlessefficient
ex-opposedpistontwo-strokeJunkersdieselengine
Closedtype:pressuredropisminimisedcomparedtoopen type

(c) Differenttypesofnozzlefordifferentcombustionchamber
(i) Singleholenozzle:
-usedinopencombustionchamber
-sizeofholelargerthan0.2mm
-veryhighinjectionpressurerequired

(ii) Multi-hole nozzle:
-no.ofholevariesfrom4 to18andthesizefrom1.5to0.35mm
-injectionrateisnotuniform

(iii) Pintlenozzle:
-aprojectionorpintleisprovidedinthenozzletoavoidweakinjectionanddribbling
-pintlemaybecylindricalorconicalshape
-coneanglevariedfrom0to60°
-providegoodatomisationandreducedpenetration
-fuelpressuresarelowerthansingleandmulti-holenozzle

(iv) Pintauxnozzle:

-injectedfuelinupstreamofair
-developmentofpintlenozzlewithauxiliaryholedrilledinthenozzlebody
-reduceddelayperiodandincreasedthermalefficiency

Module-IIIGNITION SYSTEM

BasicallyConvectionalIgnitionsystemsareof2 types:
(a) BatteryorCoilIgnitionSystem,and
(b) MagnetoIgnitionSystem

Boththeseconventional,ignitionsystemsworkonmutualelectromagneticinduction principle.

BatteryorCoilIgnitionSystem:
-usedin4-wheelers,butnow-a-daysitismorecommonlyusedin2-wheelersalso(i.e.Button start, 2-wheelers like Pulsar, Kinetic Honda; Honda-Activa, Scooty, Fiero, etc.)

(e) Theignitionsystemisdividedinto2-circuits:

(i) PrimaryCircuit:
-consistsof6or12Vbattery,ammeter,ignitionswitch,primarywinding
-ithas200-300turnsof20SWG(SharpsWireGauge)gaugewire,contactbreaker, capacitor.

(ii) SecondaryCircuit:

-consistsofsecondarywindingorcoilwhichhave21000turnsof40(SWG)gaugewire.
-bottomendofwhichisconnectedtobottomendofprimaryandtopendofsecondary winding or coil is connected to centre of distributor rotor.
-distributorrotorsrotateandmakecontactswithcontactpointsandareconnectedtospark plugs which are fitted in cylinder heads.

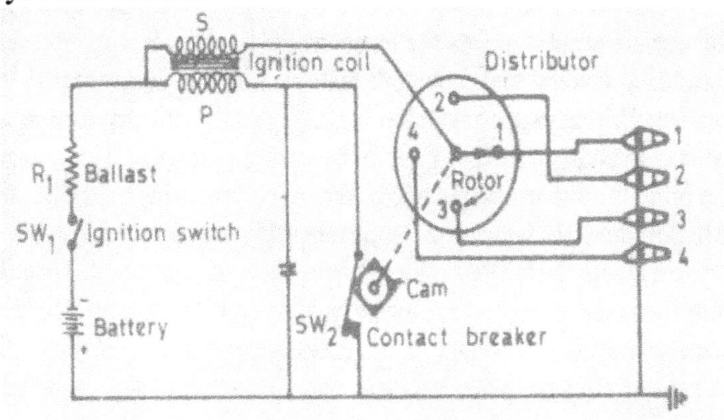

C=condenser,P=primarycoil,S=secondarycoil,R1=ballastresistance,SW1=ignitionswitch, SW2=contact breaker

Fig.22.Circuitdiagramforaconventionalsparkignitionsystem

Working:

-When the ignition switch is closed and engine in cranked, as soon as the contact breaker closes, a lowvoltage current will flowthrough the primarywinding. It is also to be noted that the contact beaker cam opens and closes the circuit 4-times (for 4 cylinders) in one revolution. When thecontact breaker opens the contact, the magnetic field begins to collapse. Because of this collapsing magnetic field, current will be induced in the secondary winding. And because of more turns (@ 21000 turns of secondary, voltage goes unto 28000-30000 volts. This high voltage current is brought to centre of the distributor rotor. Distributor rotor rotates and supplies this high voltage current to proper stark plug depending upon the engine firing order. When the high voltage current jumps the spark plug gap, it produces the spark and the charge is ignited-combustion starts-products of combustion expand and produce power.

-The Function of the capacitor is to reduce arcing at the contact breaker (CB) points. Also when the CB opens the magnetic field in the primary winding begins to collapse. When the magnetic field is collapsing capacitor gets fully charged and then its tarts discharging and helps in building up of voltage in secondary winding.
-Contactbreakercamanddistributorrotoraremountedonthesameshaft.
-In 2-stroke cycle engines these are motored at the same engine speed. And in 4-stroke cycle engines they are motored at half the engine speed.
(f) Agood sparkisavailableatlowspeedalso.
(g) Occupiesmorespace.
(h) Rechargingisamustincasebatterygetsdischarged.

MagnetoIgnitionSystem:
(i) magnetowillproduceandsupplytherequiredcurrenttotheprimarywindingorcoil
(j) rotating magneto with fixed coil or rotating coil with fixed magneto for producing and supplying current to primary, remaining arrangement is same as that of a battery ignition system
-nobatteryrequired
(k) duringstartingthequalityofsparkispoordue toslowspeed
-verymuchcompact

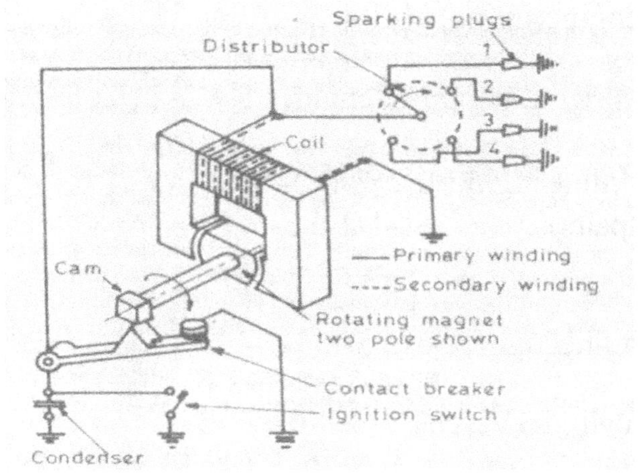

Fig.23.Hightensionmagnetoignitionsystem

Disadvantage of conventional ignition systems

Followingarethedrawbacksofconventionalignitionsystems:

(a) Becauseofarcing,pittingofcontactbreakerpoint

(b) Poorstarting:Afterfewthousandsofkilometresofrunning,thetimingbecomes inaccurate, which results into poor starting (Starting trouble).

(c) Atveryhighenginespeed,performanceispoorbecauseofinertiaeffectsofthemoving parts in the system.

(d) Sometimesitisnotpossibletoproducesparkproperlyinfouledsparkplugs.

Duetoaboveproblemelectronicignitionsystemis used.

Followingaretheadvantagesofelectronicignitionsystem:

(a) Movingpartsareabsent-sonomaintenance.

(b) Contactbreakerpointsareabsent-sonoarcing.

(c) Sparkpluglifeincreasesby50%andtheycanbeusedforabout60000kmwithoutany problem.

(d) Bettercombustionincombustionchamber,about90-95%ofairfuelmixtureisburnt compared with 70-75% with conventional ignition system.

(e) Morepoweroutput.

(f) Morefuelefficiency.

Firingorder:

Theorderorsequenceinwhichthefiringtakesplace,indifferentcylindersofamulti- cylinder engine is called Firing Order.

IncaseofSIenginesthedistributorconnectsthesparkplugsofdifferentcylindersaccording to Engine Firing Order.

Advantages

(a) Aproperfiringorderreducesenginevibrations

(b) Maintainsenginebalancing.

(c) Securesanevenflowofpower.

- Firingorderdiffersfromengine-to-engine.
- Probablefiringordersfor differentenginesare:

−3cylinder=1-3-2

−4cylinderengine(inline)=1-3-4-2
1-2-4-3

−4cylinderhorizontalopposedengine=1-4-3-2 (Volkswagen engine)

−6-cylinderinlineengine=1-5-3-6-2-4
(Cranksin 3pairs) 1-4-2-6-3-5
1-3-2-6-4-5
1-2-4-6-5-3

−8cylinderinlineengine1-6-2-5-8-3-7-4
1-4-7-3-8-5-2-6

8cylinderVtype1-5-4-8-6-3-7-2
1-5-4-2-6-3-7-8
1-6-2-5-8-3-7-4
1-8-4-3-6-5-7-2

Cylinder1istakenfromfrontofinlineandfrontrightsideinV engines.

Ignitiontiming:

Itis veryimportant,sincethechargeistobeignitedjustbefore(fewdegreesbeforeTDC)the end of compression, since when the charge is ignited, it will take some time to come to the required rate of burning.

IgnitionAdvance:

The purpose of spark advance mechanism is to assure that under every condition of engine operation,ignitiontakesplaceatthemostfavourableinstantintimei.e.mostfavourablefrom a standpoint of engine power, fuel economy and minimum exhaust dilution. By means of these mechanisms the advance angle is accurately set so that ignition occurs before TDC point of the piston. The engine speed and the engine load are the control quantities required for the automatic adjustment of the ignition timing. Most of the engines are fitted with mechanisms which are integral with the distributor and automatically regulate the optimum spark advance to account for change of speed and load. The two mechanisms used are: (a)Centrifugal advance mechanism, and
(b)Vacuumadvancemechanism

Factorsaffectingenergyrequirementofignitionsystem

(a) Effectofseries resistance:
-Energydissipatedduringdischargeandhencecircuitenergystoredincapacitanceismore
-longerdischargetime

(b) Effectofelectrodematerial:
-ignitionenergyforelectrodegapslargerthanthequenchingdistancevarieswithmaterials for the electrode and increases with any change to material having higher boiling point

(c) Effectofsprayinductance:
-presenceofsmallstrayinductancecauseoscillatorydischarge
-largerresistance,minimuminductancelongerthedischargetime

(d) Effectofelectrodeconfiguration:
-reductioninthecapacitancereducesthedischargetimeandincreaseinthegasresistance which reduces the diameter of the spark channel
-increasetheelectrodegapmeanssupplyofignitionenergyoverintervaloftimedecreases and spark channel also reduces

***StagesofSIenginecombustion**

In SI engine homogeneous mixture of vaporised fuel, air and residual gases is ignited by a single intense and high temperature spark between the spark plug electrode (electrodes exceeds 10,000 °C) and generate pre-flame which spreads to envelope of mixture for combustion.

ThreestagesofcombustioninSIengineare,

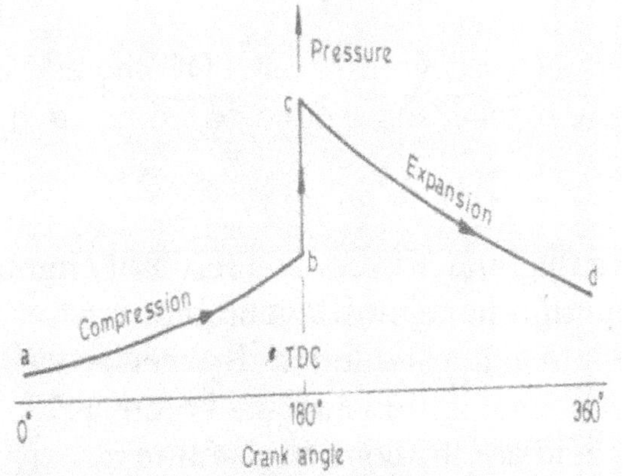

Fig.24.Theoreticalp-θdiagram

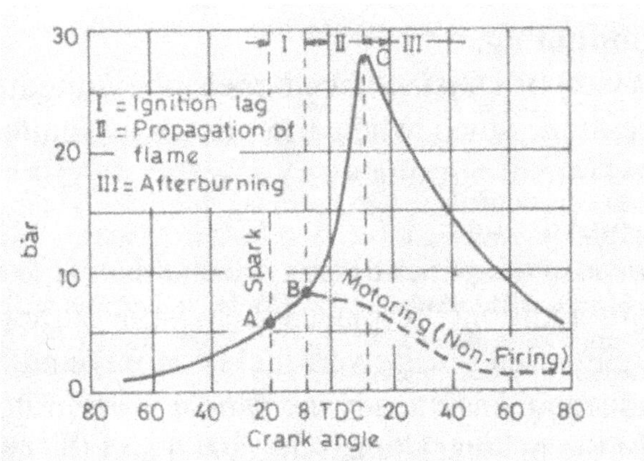

Fig.25.StagesofcombustioninSIengine

I-Ignitionlagorpreparationphase(AB):

-growthanddevelopmentofasemipropagatingnucleusofflame
-chemicalprocessdependinguponthenatureofthefuel,uponbothtemperatureandpressure, the proportion of the exhaust gas, and also upon the temperature coefficient of the fuel, thatis, the relationship of oxidation or burning
-pointAshowsthepassageofsparkandpointBisthefirstriseofpressure
-ignitionlagisgenerallyexpressedintermsofcrankangle
-Ignitionlagisverysmallandliesbetween0.00015to0.0002 seconds
-ignitionlagof0.002secondscorrespondsto35degcrankrotationwhentheengineis running at 3000 RPM
-Angleofadvanceincreasewiththespeed

II- propagationofflame(BC):

-PeriodfromthepointBwherethelineofcombustiondepartsfromthecompressionlineto point C, the maximum rise of pressure in P-θ diagram
-flamepropagatesattheconstantvelocity
-Heattransfertothecylinderwallislow
-rateofheatreleasedependsupontheturbulenceintensityandreaction rate

III- Afterburning(CD):

-AfterpointC,theheatreleaseisduetothefuelinjectioninreducedflamefrontafterthe starts of expansion stroke
-nopressureriseduringthisperiod

***EffectofenginevariablesonIgnitionlag**

Fuel:Highself-ignitiontemperatureoffuellongertheignitionlag.
Mixture ratio: mixture richer than the stoichiometric ratio provide shorter ignition lag
Initialtemperatureandpressure: increasingtheintaketemperatureandpressure,increasing thecompressionratio,chemicalreactionrateandretardingthesparkallreducetheignition lag
Electrodegap:lowerthecompressionratioandhighertheelectrodegapisdesirable
-voltagerequiredatthesparkplugelectrodetoproducesparkisfoundtoincreasewith decrease in fuel-air ratio and with increase in compression ratio and engine load
Turbulence: directly proportional to engine speed
-enginespeeddoesnotaffectmuchignitionlagmeasuredinmilliseconds
-butignitionlagincreaseslinearlywithenginespeedwhenmeasuredindegreecrankangle

-sparkadvanceisdesirableinhigherenginespeed

***Effectofenginevariablesonflamepropagation**

Rate of flame propagation affects the combustion process in SI engines. Higher combustion efficiency and fuel economy can be achieved by higher flame propagation velocities. Unfortunately flame velocities for most of fuel range between 10 to 30 m/second.

Thefactorswhichaffecttheflamepropagationsare

1. Airfuelratio
2. Compressionratio
3. Loadonengine
4. Turbulenceandenginespeed
5. Otherfactors

1. A:F ratio: The mixture strength influences the rate of combustion and amount of heat generated.Themaximumflamespeedforallhydrocarbonfuelsoccursatnearly10%rich mixture. Flame speed is reduced both for lean and as well as for very rich mixture. Lean mixturereleaseslessheatresultinglowerflametemperatureandlowerflamespeed.Veryrich mixture results incomplete combustion and also results in production of less heat and flame speed remains low.

2. Compression ratio: The higher compression ratio increasesthe pressure andtemperatureof the mixture and also decreases the concentration of residual gases. All these factors reducetheignitionlagandhelptospeedupthesecondphaseofcombustion.Themaximumpressure of the cycle as well as mean effective pressure of the cycle with increase in compressionratio. Figure above shows the effect of compression ratio on pressure (indirectlyon the speed of combustion) with respect to crank angle for same A: F ratio and same angle of advance. Higher compression ratio increases the surface to volume ratio and therebyincreases the part of the mixture which after-burns in the third phase.

3. Load on Engine: With increase in load, the cycle pressures increase and the flame speed also increases. In S.I.engine,thepowerdevelopedbyanengineiscontrolledbythrottling. At lower load and higher throttle, the initial and final pressure of the mixture after compression decrease and mixture is also diluted by the more residual gases. This reduces the flame propagation and prolongs the ignition lag. This is the reason, the advance mechanism is also provided with change in load on the engine. This difficulty can be partly overcome by providing rich mixture at part loads but this definitely increases the chances of afterburning. The after burning is prolonged with richer mixture. In fact, poor combustion at part loads and necessity of providing richer mixture are the main disadvantages of SI engines which causes wastage of fuel and discharge of large amount of CO with exhaust gases.

4. Turbulence: Turbulence plays very important role in combustion of fuel as the flame speed is directly proportional to the turbulence of the mixture. This is because, the turbulence increasesthe mixingand heat transfer coefficient or heat transfer rate betweenthe burned and unburned mixture. The turbulence of the mixture can be increased at the end of compression by suitable design of the combustion chamber (geometry of cylinder head and piston crown). Insufficient turbulence provides low flame velocity and incomplete combustion and reduces the power output. But excessive turbulence is also not desirable as it increases thecombustion rapidly

and leads to detonation. Excessive turbulence causes to cool the flame generated and flame propagation is reduced. Moderate turbulence is always desirable as it accelerates the chemical reaction, reduces ignition lag, increases flame propagation and even allows weak mixture to burn efficiently.

Engine Speed

The turbulence of the mixture increases with an increase in engine speed. For this reason the flame speed almost increases linearly with engine speed. If the engine speed is doubled,flame to traverse the combustion chamber is halved. Double the original speed and half the original time give the same number of crank degrees for flame propagation. The crank angle required for the flame propagation, which is main phase of combustion will remain almost constant at all speeds. This is an important characteristic of all petrol engines.

EngineSize

Engines of similar design generally run at the same piston speed. This is achieved by using small engines having larger RPM and larger engines having smaller RPM. Due to samepiston speed, the inlet velocity, degree of turbulence and flame speed are nearly same in similarenginesregardlessofthesize.However,insmallenginestheflametravelissmalland in large engines large. Therefore, if the engine size is doubled the time required for propagationofflamethroughcombustionspaceisalsodoubled.ButwithlowerRPMoflarge engines the time for flame propagation in terms of crank would be nearly same as in small engines. In other words, the number of crank degrees required for flame travel will be about the same irrespective of engine size provided the engines are similar.

5. Other Factors: Among the other factors, the factors which increase the flame speed are supercharging of the engine, spark timing and residual gases left in the engine at the end of exhaust stroke. The air humidity also affects the flame velocity but its exact effect is not known. Anyhow, its effect is not large compared with A:F ratio and turbulence.

*Detonationorknocking

Knocking is due to auto ignition of end portion of unburned charge in combustion chamber. As the normal flame proceeds across the chamber, pressure and temperature of

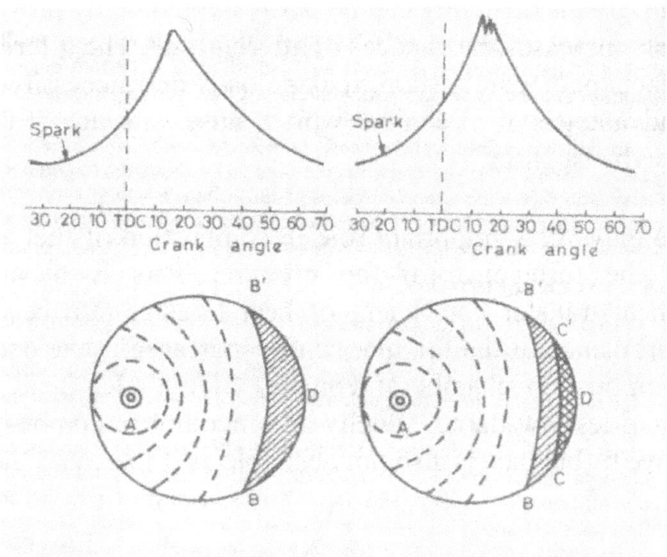

(a) Normal combustion (b) Detonation or knock

unburned charge increase due to compression by burned portion of charge. This unburned compressed charge may auto ignite under certain temperature condition and release the energy at a very rapid rate compared to normal combustion process in cylinder. This rapid release of energy during auto ignition causes a high pressure differential in combustion chamber and a high pressurewaveisreleasedfromautoignitionregion. Themotionofhighpressurecompression waves inside the cylinder causes vibration of engine parts and pinging noise and it is known asknockingor detonation.Thispressure frequencyorvibrationfrequencyinSIenginecanbe up to 5000 Cycles per second. Denotation is undesirable as it affects the engine performance and life, as it abruptly increases sudden large amount of heat energy. It also put a limit on compression ratio at which engine can be operated which directly affects the engineefficiency and output.

Fig.26.DetonationinSIengine

***Autoignition**

A mixture of fuel and air can react spontaneously and produce heat by chemical reaction in the absence of flame to initiate the combustion or self-ignition. This type of self-ignition in the absence of flame is known as Auto-Ignition. The temperature at which the self-ignition takes place is known as self-igniting temperature. The pressure and temperature abruptly increase due to auto-ignition because of sudden release of chemical energy. This auto-ignition leads to abnormal combustion known as detonation which is undesirable because its bad effect on the engine performance and life as it abruptly increases sudden large amount of heat energy. In addition to this knocking puts a limit on the compression ratio at which an engine can be operated which directly affects the engine efficiency and output.

***Pre-ignition**

Pre-ignition is the ignition of the homogeneous mixture of charge as it comes in contact with hot surfaces, in the absence of spark. Auto ignition may overheat the spark plug and exhaust valve and it remains so hot that its temperature is sufficient to ignite the charge in next cycle during the compression stroke before spark occurs and this causes the pre-ignition of the charge. Pre-ignition is initiated by some overheated projecting part such as the sparking plug electrodes, exhaust valve head, metal corners in the combustion chamber, carbon deposits or protruding cylinder head gasket rim etc. pre-ignition is also caused by persistent detonating pressure shockwaves scoring away the stagnant gases which normallyprotect the combustion chamber walls. The resulting increased heat flow through the walls, raises the surface temperature of any protruding poorly cooled part of the chamber, and this therefore providesa focal point for pre-ignition.

EffectsofPre-ignition

-Itincreasethetendencyofdenotationinthe engine
-Itincreasesheattransfertocylinderwallsbecausehightemperaturegasremainsincontact with for a longer time
-Pre-ignitioninasinglecylinderwillreducethespeedandpoweroutput
-Pre-ignitionmaycauseseizerinthemulti-cylinderengines,onlyifonlycylindershavepre-ignition

***Effectofdetonation**

Theharmfuleffectsofdetonationareasfollows:

1. *Noiseandroughness:*knockingproducesaloudpulsatingnoiseandpressurewaves.These waves vibrate back and forth across the cylinder. The presence of vibratory motion causes crankshaft vibrations and the engine runs rough.

2. *Mechanicaldamage:*

(a) High pressure waves generated during knocking can increase rate of wear of parts of combustion chamber. Sever erosion of piston crown (in a manner similar to that of marine propeller blades by cavitation), cylinder head and pitting of inlet and outlet valves may result in complete wreckage of the engine.

(b) Detonation is very dangerous in engines having high noise level. In small engines the knocking noise is easily detected and the corrective measures can be taken but in aero- enginesit isdifficult to detectknockingnoise and hencecorrective measurescannot be taken. Hence severe detonation may persist for a long time which may ultimately result in complete wreckage of the piston.

3. *Carbondeposits:*Detonationresultsinincreasedcarbon deposits.

4. *Increaseinheattransfer:*Knockingisaccompaniedbyanincreaseintherateofheat transfer to the combustion chamber walls.

Theincreaseinheattransferisduetotwo reasons.

-Theminorreasonisthatthemaximumtemperatureinadetonatingengineisabout150°C higher than in a non-detonating engine, due to rapid completion of combustion

-The major reason for increased heat transfer is the scouring away of protective layer of inactivestagnantgasonthecylinderwallsduetopressurewaves.Theinactivelayerofgasnormally reducesthe heat transfer byprotectingthe combustion and piston crown from direct contact with flame.

5. *Decrease in power output and efficiency:*Due to increase in the rate of heat transfer the power output as well as efficiency of a detonating engine decreases.

*6 Pre-ignition:*increase in the rate of heat transfer to the walls has yet another effect. It may cause local overheating, especially of the sparking plug, which may reach a temperature high enough to ignite the charge before the passage of spark, thus causing pre-ignition. An engine detonating for a long period would most probably lead to pre-ignition and this is the real danger of detonation.

***EffectofengineoperatingvariablesontheengineknockingDetonation**

Thevariousenginevariablesaffectingknockingcanbeclassifiedas:

-Temperaturefactors

-Densityfactors

-Timefactors

-Compositionfactors

(a) Temperaturefactors:

Increasing the temperature of the unburned mixture increase the possibility of knock in SI engine, the effect of following engine parameters on the temperature of the unburnedmixture:

-Raising the compression ratio: Increasing the compression ratio increases both the temperature and pressure (density of the unburned mixture). Increase in temperature

reduces the delay period of the end gas which in turn increases the tendency to knock.
-Supercharging: It also increases both temperature and density, which increase the knocking tendency of engine
-Coolant temperature: Delay period decreases with increase of coolant temperature,decreased delay period increase the tendency to knock
-Temperature of the cylinder and combustion chamber walls: The temperature of the end gas depends on the design of combustion chamber. Sparking plug and exhaust valve are two hottest parts in the combustion chamber and uneven temperature leads to pre-ignition and hence the knocking.

(b) Densityfactors:
Increasing the density of unburnt mixture will increase the possibility of knock in the engine. The engine parameters which affect the density are as follows:
-Increasedcompressionratioincreasethedensity
-Increasingtheloadopensthethrottlevalvemoreandthusthedensity
-Superchargingincreasethedensityofthe mixture
-Increasing the inlet pressure increases the overall pressure during the cycle. The high pressure end gas decreases the delay period which increase the tendency of knocking.
-Advanced spark timing: quantity of fuel burnt per cycle before and after TDC position depends on spark timing. The temperature of charge increases by increasing the spark advance and it increases with rate of burning and does not allow sufficient time to the end mixture to dissipate the heat and increase the knocking tendency

(c) Timefactors:
Increasingthetimeofexposureoftheunburnedmixtureto auto-ignitionconditionsincrease the possibility of knock in SI engines.

Flame travel distance: If the distance of flame travel is more, then possibility of knocking is also more. This problem can be solved by combustion chamber design, spark plug location and engine size. Compact combustion chamber will have better anti-knock characteristics, since the flame travel and combustion time will be shorter. Further, if the combustion chamber is highlyturbulent, the combustion rate ishigh and consequentlycombustion time is further reduced; this further reduces the tendency to knock.
Location of sparkplug: A spark plug which is centrally located in the combustion chamberhas minimum tendency to knock as the flame travel is minimum. The flame travel can be reduced by using two or more spark plugs.
Locationofexhaustvalve: Theexhaustvalveshouldbelocatedclosetothesparkplugsothat it is not in the end gas region; otherwise there will be a tendency to knock.
Engine size: Large engines have a greaterknockingtendencybecause flame requires alonger time to travel across the combustion chamber. In SI engine therefore, generally limited to 100mm
Turbulence of mixture: decreasing the turbulence of the mixture decreases the flame speed and hence increases the tendency to knock. Turbulence depends on the design of combustion chamber and one engine speed.

(d) Composition:

ThepropertiesoffuelandA/Fratioareprimarymeanstocontrolknock:

(i) Molecular Structure: The knocking tendency is markedly affected by the type of the fuel used. Petroleum fuels usuallyconsist of many hydro-carbons of different molecular structure. The structure of the fuel molecule has enormous effect on knocking tendency. Increasing the carbon-chain increases the knocking tendency and centralizing the carbon atoms decreasesthe knocking tendency. Unsaturated hydrocarbons have less knocking tendency thansaturated hydrocarbons.

Paraffins

-Increasingthelengthofcarbonchainincreasestheknockingtendency.

-Centralisingthecarbonatomsdecreasestheknockingtendency.

-Addingmethyl group(CHtothesideofthecarbonchaininthecentrepositiondecreasesthe knocking tendency.

Olefins

-Introductionofonedoublebondhaslittleeffectonanti-knockqualitybuttwo orthree double bond results less knocking tendency except C and C

Napthenesand Aromatics

-Naptheneshavegreaterknockingtendencythancorrespondingaromatics.

-Withincreasingdouble-bonds,theknockingtendencyisreduced.

-Lengtheningthesidechainsincreasestheknockingtendencywhereasbranchingoftheside chain decreases the knocking tendency.

(ii) Fuel-airratio:Themostimportanteffectoffuel-aftratioisonthereactiontimeorignition delay. Whenthemixtureisnearly10%richerthanstoichiomiric(fuel-airratio=0.08)ignition lagofthe end gas is minimum andthevelocityof flamepropagationis maximum. Bymaking the mixture leaner or richer (than F/A 0.08) the tendency to knock is decreased. A too rich mixture is especially effective in decreasing or eliminating the knock due to longer delay and lower temperature of compression.

(iii)Humidity of air: Increasing atmospheric humidity decreases the tendency to knock by decreasing the reaction time of the fuel

*SI engine combustion chamber

(a) T-headcombustionchamber:2camshaft,pronetodetonation,averageoctanenumber 45- 50

(b) I-headorsidevalvecombustion chamber:

-lackofturbulence

-extremelysensitivetoignitiontiming

-pronetodetonation

(c) F-headcombustionchamber:

-CompromisebetweenI-headandL-head

*StagesofcombustioninCIengine

1. Ignitiondelayperiod:

The period between the start of fuel injection into the combustion chamber and the start of combustion is termed as ignition delay period. The start of combustion is determined fromthe change in slope on p-θ diagram or from heat release analysis of the p-θ data, or from luminosity detector in experimental conditions. Start of injection can be determined by a needle-lift indicator to record the time when injector needle lifts off its seat. Start of combustion is more difficult to determine precisely. It is best identified from the change in slope of heat release rate, determined from cylinder pressure data. In DI engines ignition is well defined, in IDI engines ignition point is harder to identify

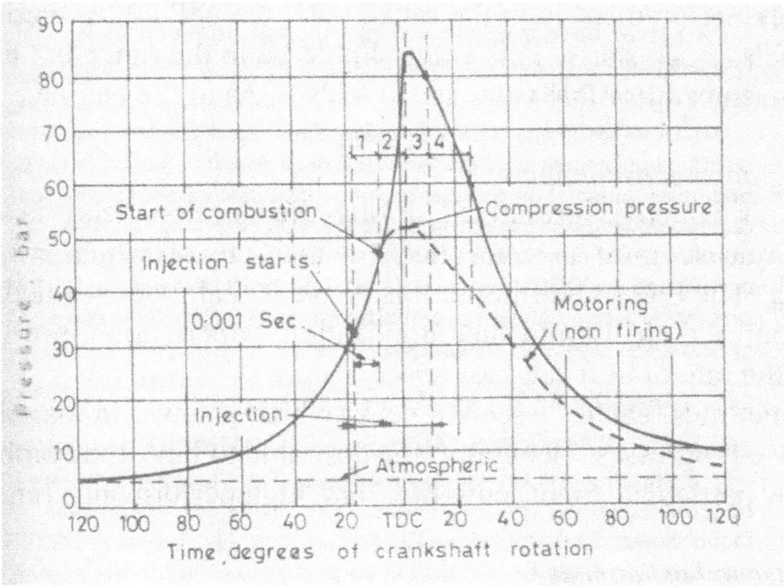

Fig.27.StagesofcombustioninCIengine

Both physical and chemical processes must take place before a significant fraction of the chemical energy of the injected liquid is released.

<u>*Physical processes*</u>are fuel spray atomization, evaporation and mixing of fuel vapour with cylinder air.Good atomization requires high fuel-injection pressure, small injector hole, optimum fuel viscosity, high cylinder pressure (large divergence angle).

Rateofvaporizationofthefueldropletsdependsondropletdiameter,velocity,fuelvolatility, pressure and temperature of the air.

<u>*Chemical processes*</u>similar to that described for auto ignition phenomenon in premixed fuel-air,onlymorecomplexsince **heterogeneousreactions** (reactionsoccurringontheliquidfuel drop surface) also occur.

Chemicaldelayismore effectivefortheduratioonoftheignitiondelayperiod. Ignition delay period is in the range of
0.6to3msforlow-compressionratioDIdieselengines,
0.4to1msforhigh-compressionratio,turbochargedDIdiesel engines,
0.6to1.5msfor IDIdieselengines

2. Rapidoruncontrolledorpre-mixedcombustionphase:

Combustion of the fuel which has mixed with air within flammability limits during ignition delay period occurs rapidly in a few crank angle degrees - high heat release characteristics in this phase. If the amount of fuel collected in the combustion chamber during the ignitiondelay is much - high heat release rate results in a rapid pressure rise which causes the diesel knock.

For fuels with low cetane number, with long ignition delay, ignition occurs late in the expansion stroke - incomplete combustion, reduced power output, poor fuel conversion efficiency. If the pressure gradient is in the range 0.4 - 0.5 MPa/°CA, engine operation is not smooth and diesels knock starts. This value should be in the range 0.2 to 0.3 MPa/°CA for smooth operation (max allowable value is 1.0 MPa/°CA) of the engine.

3. Controlledordiffusioncombustionphase:

Once the fuel and air which is pre-mixed during the ignition delay is consumed, the burning rate (heat release rate) is controlled by the rate at which mixture becomes available for burning. The rate of burning in this phase is mainly controlled by the mixing process of fuel vapour and air. Liquid fuel atomization, vaporization, pre flame chemical reactions alsoaffect the rate of heat release.

Heat release rate sometimes reaches a second peak (which is lower in magnitude) and then decreases as the phase progresses. Generally it is desirable to have the combustion process near the TDC for low particulate (soot) emissions and high performance (and efficiency).

4. Afterburningorlatecombustionphase:

Heat release rate continues at a lower rate into the expansion stroke -thereare several reasons for this: a small fraction of the fuel may not yet burn, a fraction of the energy is present in soot and fuel-rich combustion products and can be released. The cylinder charge is non- uniform and mixing during this phase promotes more complete combustion and less dissociated product gases. Kinetics is slower.

➢ **Variablesaffectingdelay period**

(i) Cetanenumber

Bothphysicalandchemicalpropertiesofthefuelareimportant.Ignitionqualityofthefuelis defined by its cetane number. Straight chain parafinic compounds (normal alkanes) havehighest ignition quality, which improves as the chain length increases. Aromatic compounds, alcohols have poor ignition quality.

-Cetane number can be increased by ignition-accelerating additives like organic peroxides, nitrates, nitrites and various sulphur compounds. Most important (commercially) is alkyl nitrates – about 0.5% by vol in a distillate fuel increase CN by 10.

-NormaldieselfuelhasCNof40to55(highspeed50–60,lowspeed25 –45

(ii) Injectiontiming

-At normal operatingconditions min ignition delay(ID) occurs with start ofinjection at 10to 15 °CA BTDC

-Cylindertemperatureandpressuredropsifinjectionisearlierorlater(highatfirstbut decrease as delay proceeds)

(iii) Injectionquantity(load)

-ReducingengineloadchangesAFR,coolsdown theengine,reduceswalltemperatures, reduces residual gas temperatures and increase ID

-Dropletsize,injectionvelocityandrateIgnitionqualitywithinpracticallimitsdonothave

significant effect on ID
-IncreaseininjectionpressureproducesonlymodestdecreaseinIDInjectornozzlediameter
-effectsofdropletsizebuthasnosignificanteffectonID

(iv) Intakeairtemperatureandpressure
-ReducingintakeairTandpincrease ID
-StrongdependenceofIDonchargetemperaturebelow1000K –abovethisvalueeffectof intake air conditions is not significant.

(v) Enginespeed
Increaseinenginespeedincreasestheairmotionandturbulence,reducesIDtimeslightly(in ms), in terms of CA degrees ID increases almost linearly.
-Achangeinenginespeed,changes“temp~time”and“pressure~time”relationships

(vi) Combustionchamberdesign
-SprayimpingementonthewallseffectfuelevaporationandIDincreaseincompression ratio, increase p and T and reduces ID
-Reducingstrokevolume,increasesurfaceareato volumeratio,increaseenginecoolingand increase ID

(vii) Swirlrate
-Changeofevaporationrateandair-fuelmixing-undernormaloperatingconditionsthe effect is small.
-Atstart-up(lowenginespeedandtemperature)moreimportant,highrateofevaporationand mixing is obtained by swirl

(viii) Oxygenconcentration
ResidualgasesreduceO_2concentrationandreducingoxygenconcentrationincreasesID

<u>*Dieselknock</u>
-CIenginedetonationoccursinthebeginningofcombustion
-InCIenginethefuelandairareimperfectlymixedandhencetherateofpressureriseis normally cause audible knock. Rate of pressure rise may reach as high as 10 bar/°CA-Highenginevibrationisthesymptomsofknocking

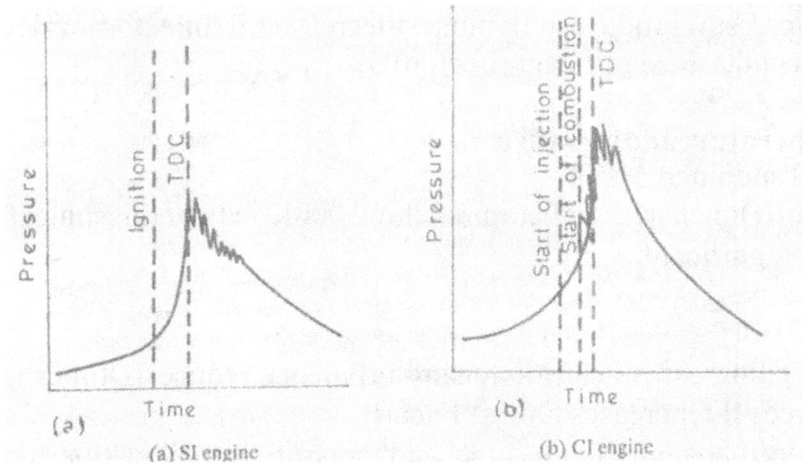

(a) SI engine (b) CI engine

-nopre-ignitionorprematureignitionaslikeSIengine

Fig.28.DetonationinSIandCIengine

<u>*Combustionchamber</u>

(i) Accordingtotheswirlofair

(a) Inductionswirloropenchamberornon-turbulentchamber
(b) Compressionswirl
(c) Combustioninducedswirl

(ii) Accordingtospeedoftheengine

(a) Lowspeedengine(n<1500rpm):

(l) Shallow,swirl-lesscombustionchambers
-Direct,multi-jetfuelinjection

(b) Mediumspeedengine(n=1500-3000rpm):
-Deepcombustionchamberswithintensiveswirlofcharge
-Directinjectionofatomisedfuel
(c) highspeedengine(n=3000-5000rpm):
-Pre-chambers(sectionalcombustionchambers)
-Indirectinjectionoffuelintothepre-chamber

Module-III

TESTING AND PERFORMANCE

Engine performance is an indication of the degree of success of the engine performs its assigned task, i.e. the conversion of the chemical energy contained in the fuel into the useful mechanical work. The performance of an engine is evaluated on the basis of the following;

(a) SpecificFuelConsumption.
(b) BrakeMeanEffectivePressure.
(c) SpecificPowerOutput.
(d) SpecificWeight.
(e) ExhaustSmokeandOtherEmissions.

Basicmeasurements:

Thebasicmeasurementstobeundertakentoevaluatetheperformanceofanengineonalmost all tests are the following:

(a) Speed
(b) Fuelconsumption
(c) Airconsumption
(d) Smokedensity
(e) Brakehorse-power
(f) Indicatedhorsepowerandfrictionhorse power
(g) HeatbalancesheetorperformanceofSIandCIengine
(h) Exhaustgasanalysis

1. Measurementofspeed:

-One of the basic measurements is that of speed. A wide variety of speed measuring devices areavailableinthe market.Theyrange fromamechanicaltachometertodigitalandtriggered electrical tachometers.

-The best method of measuring speed is to count the number of revolutions in a given time. This gives an accurate measurement of speed. Many engines are fitted with such revolution counters.

-A mechanical tachometer or an electrical tachometer can also be used for measuring the speed.

-The electrical tachometer has a three-phase permanent-magnet alternator to which a voltmeter is attached. -The output of the alternator is a linear function of the speed and is directly indicated on the voltmeter dial.

-Both electrical and mechanical types of tachometers are affected by the temperature variations and are not very accurate. For accurate and continuous measurement of speed a magnetic pick-up placed near a toothed wheel coupled to the engine shaft can be used.

-The magnetic pick-up will produce a pulse for every revolution and a pulse counter will accurately measure the speed.

2. Fuelconsumptionmeasurement:

Fuelconsumptionismeasuredintwo ways:

(a) The fuel consumption of an engine is measured by determining the volume flow in a given time interval and multiplying it by the specific gravity of the fuel which should be measured occasionally to get an accurate value.
(b) Anothermethodistomeasurethetimerequiredforconsumptionofagivenmassof fuel.

Asalreadymentionedtwobasictypesoffuelmeasurementmethods are:
-Volumetrictype
-Gravimetrictype
VolumetrictypeflowmeterincludesBurettemethod,AutomaticBuretteflowmeterand Turbine flow meter.

GravimetricFuelFlowMeasurement

The efficiency of an engine is related to the kilograms of fuel which are consumed and notthenumberoflitres.Themethodofmeasuringvolumeflowand thencorrectingitforspecific gravity variations is quite inconvenient and inherently limited in accuracy. Instead if the weight of the fuel consumed is directly measured a great improvement in accuracy and cost can be obtained. There are three types of gravimetric type systems which are commercially available include Actual weighing of fuel consumed, Four Orifice Flow meter, etc.

3. Measurementofairconsumption:

In IC engines, the satisfactory measurement of air consumption is quite difficult because the flow is pulsating, due to the cyclic nature of the engine and because the air a compressible fluid.Therefore,thesimplemethodofusinganorificeintheinductionpipeisnotsatisfactory since the reading will be pulsating and unreliable.

All kinetic flow-inferring systems such as nozzles, orifices and venturies have a square law relationship between flow rate and differential pressure which gives rise to severe errors on unsteady flow. Pulsation produced errors are roughly inversely proportional to the pressure across the orifice for a given set of flow conditions. The various methods and meters used for air flow measurement include,
(a) Airboxmethod,and
(b) Viscous-flowairmeter

4. Measurementofbrakepower:

Thebrakepowermeasurementinvolvesthedeterminationofthetorqueandtheangularspeed of the engine output shaft. The torque measuring device is called a dynamometer.

Dynamometers can be broadly classified into two main types, power absorption dynamometers and transmission dynamometer.

AbsorptionDynamometers

These dynamometers measure and absorb the power output of the engine to which they are coupled. The power absorbed is usually dissipated as heat by some means. Example of such dynamometers is prony brake, rope brake, hydraulic dynamometer, etc.

TransmissionDynamometers

In transmission dynamometers, the power is transmitted to the load coupled to the engine after it is indicated on some type of scale. These are also called torque-meters.

(a) Pronybrakedynamometer

One of the simplest methods of measuring brake power (output) is to attempt to stop the engine bymeansofa brake on the flywheel and measuretheweight which anarm attached to the brake will support, as it tries to rotate with the flywheel.

It consists of wooden block mounted on a flexible rope or band the wooden block when pressed into contact with the rotating drum takes the engine torque and the power isdissipated in frictional resistance. Spring-loaded bolts are provided to tighten the wooden block and hence increase the friction.

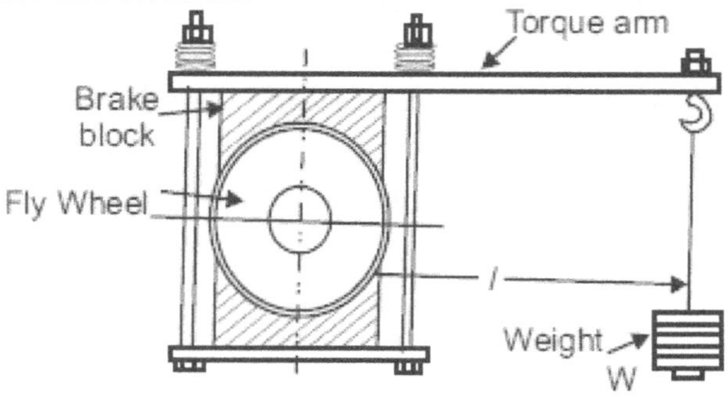

Fig.29.Pronybrake

The whole of the power absorbed is converted into heat and hence this type of dynamometer must the cooled. The brake horsepower is given by

$$BP = 2\text{p}\, NT$$

where, $T = W \times l$ W being the weight applied at a radius l.

(b) Rope brake

It consists of a number of turns of rope wound around the rotating drum attached to theoutput shaft. One side of the rope is connected to a spring balance and the other to a loading device. The power is absorbed in friction between the rope and the drum. The

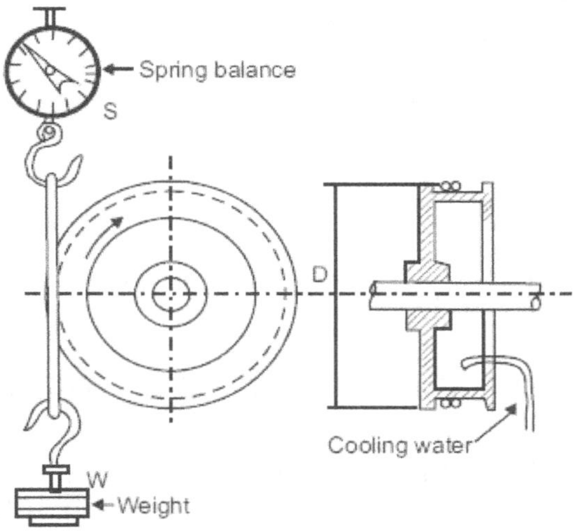

drum therefore requires cooling.

Fig.30.Ropebrake

Ropebrakeischeapandeasilyconstructedbutnotaveryaccuratemethodbecauseof changes in the friction coefficient of the rope with temperature.

The*bp*isgiven by

$$bp = \text{p}DN\,(W-S)$$

where,*D*isthebrakedrumdiameter, *W*istheweightinNewtonand *S*isthespringscale reading.

(c) Hydraulic Dynamometer

Hydraulicdynamometerworksontheprincipleofdissipatingthepowerinfluidfriction rather than in dry friction.

-Inprincipleitsconstructionissimilartothatofafluid flywheel.

-Itconsistsofaninnerrotatingmemberorimpellercoupledtotheoutputshaftofthe engine.

-Thisimpellerrotatesinacasingfilledwith fluid.

-Thisoutercasing,duetothecentrifugalforcedeveloped,tendstorevolvewiththeimpeller, but is resisted by a torque arm supporting the balance weight.

-Thefrictionalforcesbetweentheimpellerandthefluidaremeasuredbythespring-balance fitted on the casing.

-Theheatdevelopeddueto dissipationofpoweriscarriedawaybyacontinuoussupplyof the working fluid, usually water.

-Theoutputcanbecontrolledbyregulatingthesluicegateswhichcanbemovedinandoutto partially or wholly obstruct the flow of water between impeller, and the casing.

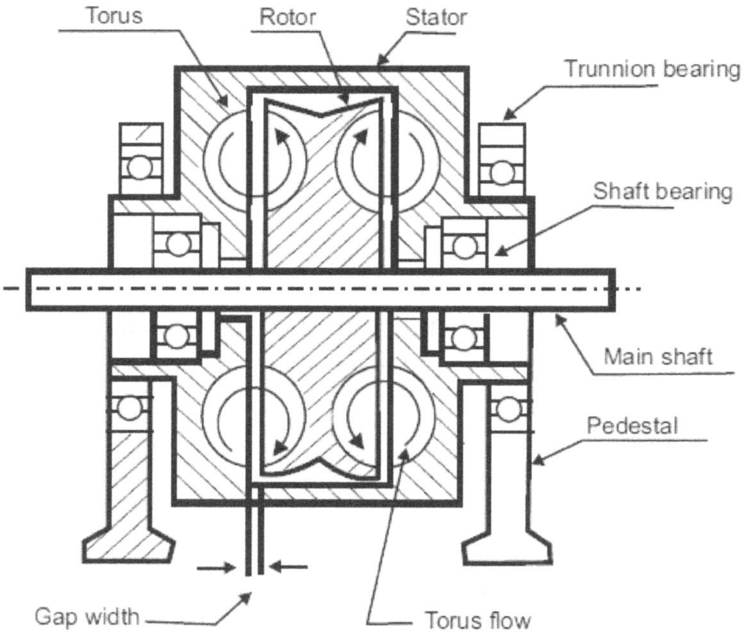

Fig.31.Hydraulicdynamometer

(d) Eddy Current Dynamometer

It consists ofa stator on which are fitted a number ofelectromagnets and a rotor disc made of copper or steel and coupled to the output shaft of the engine. When the rotor rotates eddy currents are produced in the stator due to magnetic flux set up by the passage of field current in the electromagnets. These eddy currents are dissipated in producing heat so that this typeof dynamometer also requires some cooling arrangement. The torque is measured exactly as in other types of absorption dynamometers, i.e. with the help of a moment arm. The load is controlled by regulating the current in the electromagnets.

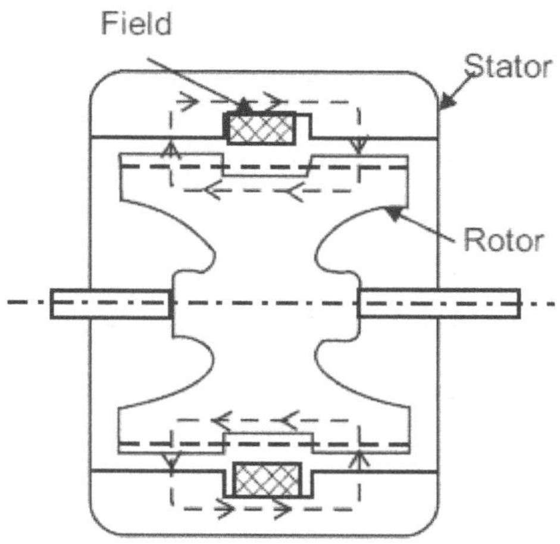

Fig.32.Eddycurrentdynamometer

Thefollowingarethemainadvantagesofeddycurrentdynamometer:

(a) Highbrakepowerperunitweightof dynamometer.

(b) Theyofferthehighestratioofconstantpowerspeedrange(upto5:1).

(c) Leveloffieldexcitationisbelow1%oftotalpowerbeinghandledbydynamometer, thus, easy to control and programme.

(d) Developmentofeddycurrentissmoothhencethetorqueisalsosmoothandcontinuous under all conditions.

(e) Relativelyhighertorqueunderlowspeedconditions.

(f) Ithasnointricaterotatingpartsexceptshaftbearing.

(g) Nonaturallimittosize-eithersmallorlarge.

(e)SwingingFieldd.c.Dynamometer

Basically, a swinging field d.c. dynamometer is a d.c. shunt motor so supported on trunnion bearings to measure their action torque that the outer case and filed coils tend to rotate with the magnetic drag. Hence, the name swinging field. The torque is measured with an arm and weighing equipment in the usual manner.

Many dynamometers are provided with suitable electric connections to run as motor also. Then the dynamometer is reversible, i.e. works as motoring as well as power absorbing device.

-When used as an absorption dynamometer it works as a d.c. generator and converts mechanical energy into electric energy which is dissipated in an external resistor or fed back to the mains.

-When used as a motoring device an external source of d.c. voltage is needed to drive the motor.

Theloadiscontrolledbychangingthefieldcurrent.

Fan Dynamometer

It is also an absorption type of dynamometer in that when driven by the engine it absorbs the engine power. Such dynamometers are useful mainly for rough testing and running. The accuracy of the fan dynamometer is very poor. The power absorbed is determined by using previous calibration of the fan brake.

Transmission Dynamometers

Transmission dynamometers, also called torque meters, mostly consist of a set of strain-gauges fixed on the rotating shaft and the torque is measured by the angular deformation of the shaft which is indicated as strain of the strain gauge. Usually, a four arm bridge is used to reduce the effect of temperature to minimum and the gauges are arranged in pairs such that the effect of axial or transverse load on the strain gauges is avoided.

Fig.33.Transmissiondynamometer

Transmissiondynamometers areveryaccurateandareusedwherecontinuoustransmissionof

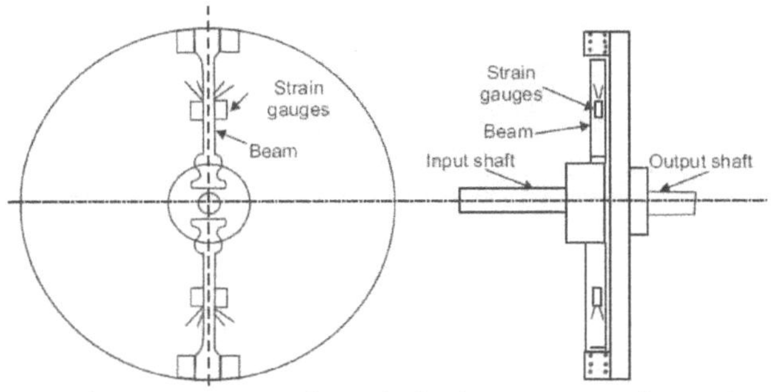

load is necessary. These are used mainly in automatic units.

5. Measurementoffrictionpower:

-The difference between indicated power and the brake power output of an engine is the friction power.

-Almost invariably, the difference between a good engine and a bad engine is due to difference between their frictional losses.

-The frictional losses are ultimately dissipated to the cooling system (and exhaust) as they appear in the form of frictional heat and this influences the cooling capacity required. Moreover, lower friction means availability of more brake power; hence brake specific fuel consumption is lower.

-The *bsfc* rises with an increase in speed. Thus, the level of friction decides the maximum output of the engine which can be obtained economically.

Thefrictionforcepowerofanengineisdeterminedbythefollowingmethods:

(a) Willan"sline method.
(b) Morsetest.
(c) Motoringtest.
(d) Differencebetween*ip*and*bp*.

(a) Willan'slinemethod

-In this method, gross fuel consumption vs. *bp* at a constant speed is plotted and the graph is extrapolated back to zero fuel consumption.

-The point where this graph cuts the *bp* axis is an indication of the friction power of the engine at that speed. This negative work represents the combined loss due to mechanical friction, pumping and blow by.

-The main drawback of this method is the long distance to be extrapolated from data measured between 5 and 40% load towards the zero line of fuel input.

-The directional margin of error is rather wide because of the graph which may not be a straight line many times.

-The changing slope along the curve indicates part efficiencies of increments of fuel. The pronounced change in the slope of this line near full load reflects the limitinginfluence ofthe air-fuel ratio and of the quality of combustion.
-Similarly, there is a slight curvature at light loads. This is perhaps due to difficulty in injecting accurately and consistently very small quantities of fuel per cycle.
-Therefore, it is essential that great care should be taken at light loads to establish the true nature of the curve.
-The Willan"s line for a swirl-chamber CI engine is straighter than that for a direct injection type engine.

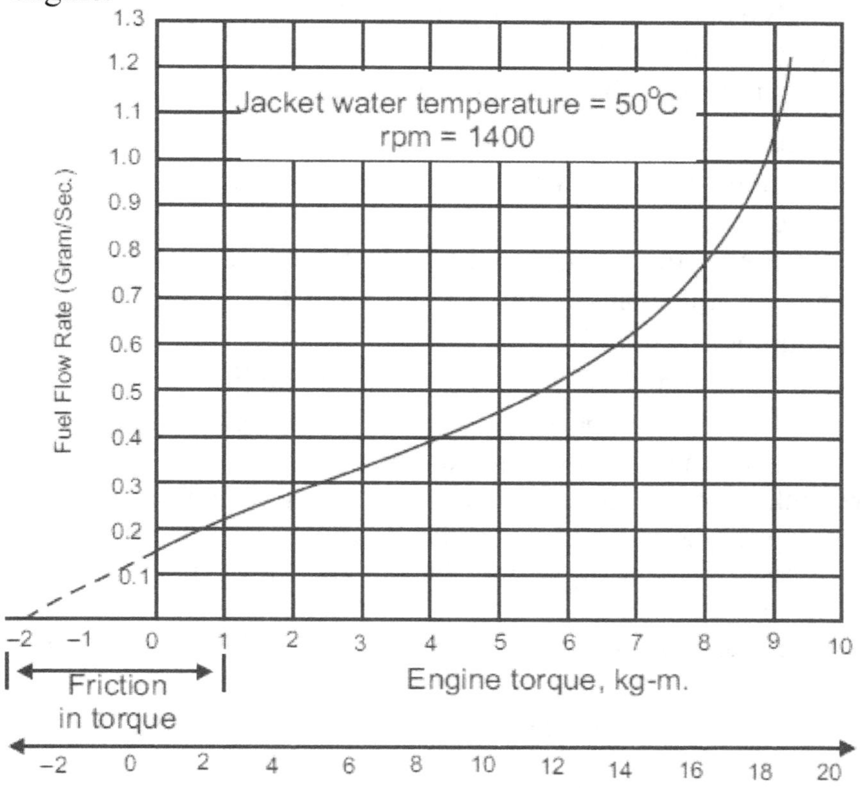

Fig.34.Willian'slinemethod

The accuracyobtained in this method is good and compares favourablywith other methods if extrapolation is carefully done.

(b) Morse Test

-TheMorsetestisapplicableonlytomulticylinderengines.
-Inthistest,theengineisfirstrunattherequiredspeedandtheoutputismeasured.
-Then,onecylinderiscutoutbyshortcircuitingthesparkplugorbydisconnectingthe injector as the case may be.
-Underthisconditionallothercylinders„motor"thiscut-outcylinder.
-Theoutputismeasuredbykeepingthespeedconstantatitsoriginalvalue.
-The difference in the outputs is a measure of the indicated horse power of the cut-out cylinder.
-Thus, for each cylinder the *ip* is obtained and is added together to find the total *ip* of the engine.
-This method though gives reasonably accurate results and is liable to errors due to changesin mixture distribution and other conditions by cutting-out one cylinder. In gasoline engines, where there is a common manifold for two or more cylinders the mixture

distribution as well as the volumetric efficiency both change. Again, almost all engines have a common exhaust manifold for all cylinders and cutting out of one cylinder may greatly affect the pulsations in exhaust system which may significantly change the engine performance by imposingdifferent back pressures.

(c) MotoringTest

-In the motoring test, the engine is first run up to the desired speed by its own power and allowed to remain at the given speed and load conditionsfor some time so that oil, water,and engine component temperatures reach stable conditions.

-The power of the engine during this period is absorbed by a swinging field type electric dynamometer, which is most suitable for this test.

-Thefuelsupplyisthencut-offandbysuitableelectric-switchingdevicesthe dynamometeris converted to run as a motor to drive for „motor" the engine at the same speed at which it was previously running.

-The power supply to the motor is measured which is a measure of the *fhp* of the engine. During the motoring test the water supply is also cut-off so that the actual operating temperatures are maintained.

-This method, though determines the *fp* at temperature conditions very near to the actual operating temperatures at the test speed and load, does, not give the true losses occurring under firing conditions due to the following reasons.

(i) Thetemperatures inthe motoredengine are different from thoseina firingengine because even if water circulation is stopped the incoming air cools the cylinder. This reduces the lubricating oil temperature and increases friction increasing the oil viscosity. This problem is much more severing in air-cooled engines.

(ii) The pressure on the bearings and piston rings is lower than the firing pressure. Load on main and connecting road bearings are lower.

(iii) The clearance between piston and cylinder wall is more (due to cooling). This reduces the piston friction.

(iv) The air is drawn at a temperature less than when the engine is firing because it does not get heat from the cylinder (rather losesheat to the cylinder). Thismakesthe expansion line to be lower than the compression line on the p-v diagram. This loss is however counted in the indicator diagram.

(v) During exhaust the back pressure is more because under motoring conditions sufficient pressure difference is not available to impart gases the kinetic energy is necessary to expel them from exhaust.

Motoring method, however, gives reasonably good results and is very suitable for finding the losses due to various engine components. This insight into the losses caused by various components and other parameters is obtained by progressive stripping-off of the under progressive dismantling conditions keeping water and oil circulation intact. Then the cylinder head can be removed to evaluate, by difference, the compression loss. In this manner piston ring, piston etc. can be removed and evaluated for their effect on overall friction.

(d) Difference between ip and bp

(i) Themethodoffindingthe *fp*bycomputingthedifferencebetween*ip*,asobtainedfroman indicator

diagram, and *bp*, as obtained by a dynamometer, is the ideal method.
(ii) Inobtainingaccurateindicatordiagrams,especiallyathighenginespeeds,thismethodis usually only used in research laboratories. Its use at commercial level is verylimited.

6. Heatbalancesheet:

Theperformanceofanengineisusuallystudiedbyheatbalance-sheet.Themaincomponents of the heat balance are:
-Heatequivalenttotheeffective(brake)workofthe engine,
-Heatrejectedtothecoolingmedium,
-Heatcarriedawayfromtheenginewiththeexhaustgases,and
-Unaccountedlosses.
The unaccounted losses include the radiation losses from the various parts of the engine and heat lost due to incomplete combustion. The friction loss is not shown as a separate item to the heat balance-sheet as the friction loss ultimately reappears as heat in cooling water, exhaust and radiation.

(i) PerformanceofSIengine:

-At full throttle the brake thermal efficiency at various speeds varies from 20 to 27 percent, maximum efficiency being at the middle speed range.

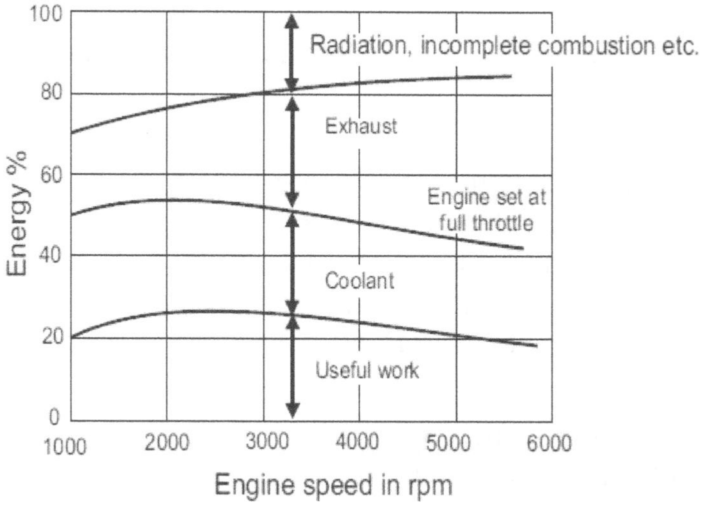

Fig.35.HeatbalancesheetforSIengine

-Thepercentageheatrejectedtocoolantismoreatlower speed(»35%) and reducesat higher speeds (» 25%). Considerably more heat is carried by exhaust at higher speeds.
-Torque and mean effective pressure do not strongly depend on the speed of the engine, but depend on the volumetric efficiency and friction losses. Maximum torque position corresponds with the maximum air charge or minimum volumetric efficiency position.
-High power arises from the high speed. In the speed range before the maximum power is obtained, doubling the speed doubles the power.
-At low engine speed the friction power is relatively low and *bhp* is nearly as large as *ip.* As engine speed increases, however, *fp* increases at continuously greater rate and therefore *bp* reaches a peak and starts reducing even though *ip* is rising. At engine speeds above the usual operating range, *fp* increases very rapidly. Also, at these higher speeds *ip* will reach a maximum and then fall off. At some point, *ip* and *fp* will be equal, and *bp* will then drop to zero.

(ii) **Performance of CI engine:**

TheperformanceofaCIengineatconstantspeedandvariableloadisgivenas,
AstheefficiencyofCIengineismorethantheSIenginethetotallossesareless.Thecoolant loss is more at low loads and radiation, etc. losses are more at high loads.

-The*bmep*,*bp*andtorquedirectlyincreasewithload

-Thelowestbrakespecificfuel consumptionandhencethemaximumefficiencyoccursat about 80% of the full load.

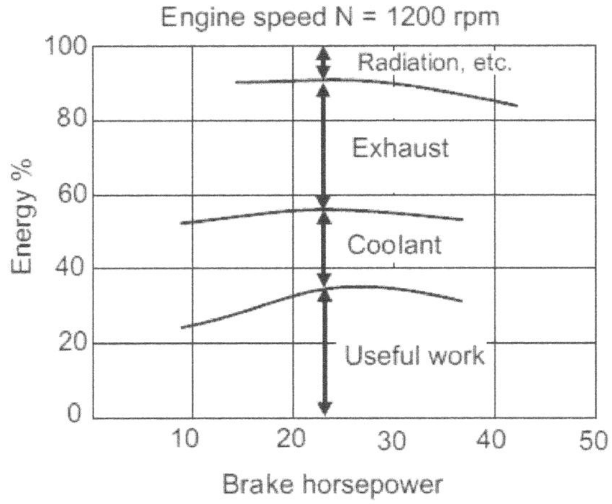

Fig.36.HeatbalancesheetofCIengineatconstantspeed

ENGINE COOLING

InaSIengine,coolingmust be satisfactorytoavoid pre-ignitionandknock. Ina compression ignition engine, since a normal combustion is aided, cooling must be sufficient to allow the parts to operate properly. In short, cooling is a matter of equalization of internal temperature to prevent local overheating as well as to remove sufficient heat energy to maintain apractical overall working temperature.

***RequirementsofcoolingsystemintheICengine**

ThecoolingsystemisprovidedintheICenginefor thefollowingreasons:

-The temperature of the burning gases in the engine cylinder reaches up to 1500 to 2000°C, which is above the melting point of the material of the cylinder body and head of the engine. (Platinum, a metal which has one of the highest melting points, melts at 1750 °C, iron at 1530°C and aluminium at 657°C.) Therefore, if the heat is not dissipated, it would result in the failure of the cylinder material.

-Due to very high temperatures, the film of the lubricating oil will get oxidized, thus producing carbon deposits on the surface. This will result in piston seizure.

-Due to overheating, large temperature differences may lead to a distortion of the engine components due to the thermal stresses set up. This makes it necessary for, the temperature variation to be kept to a minimum.

-Highertemperaturesalsolowerthevolumetricefficiencyoftheengine.

Effectofovercooling:

-Thermalefficiencyisdecreasedduetomorelossofheatcarriedbythecoolant
-Thevaporizationofthefuelislessresultinginlowercombustionefficiency
-Lowtemperatureincreasestheviscosityoflubricantcausingmorelossduetofriction

Therearemainlytwotypesofcooling systems:

(a) Aircooledsystem,and
(b) Watercooledsystem

AirCooledSystem:

-Air cooled system is generally used in small engines say up to 15-20 kW and in aero plane engines.
-In this system fins or extended surfaces are provided on the cylinder walls, cylinder head, etc.Heatgeneratedduetocombustionintheenginecylinderwillbeconductedto thefinsand when the air flows over the fins, heat will be dissipated to air.

-Theamountofheatdissipatedtoairdependsupon:

(a) Amountofairflowingthroughthefins
(b) Finsurfacearea
(c) Thermalconductivityofmetalusedforfins

-Forefficientcoolingthelengthofthefinsandthespacingbetweenthemisquiteimportant
-Largerinterspacingbetweenthefinsofferslargerareaforcoolingairbuttheheatingofthe air is less, so more cooling air is required
-Smallerinterspacingbetweenthefins resultsinsmallerflowareaofcoolingairandhence input cooling air is less
-Usuallyfinheightvariesfrom15to25mm

Advantagesofaircooledengines

Aircooledengineshavethefollowingadvantages:

1. Itsdesignofair-cooledengineis simple.
2. Itislighterinweightthanwater-cooledenginesduetotheabsenceofwaterjackets, radiator, circulating pump and the weight of the cooling water.
3. Itischeapertomanufacture.
4. Itneedslesscareandmaintenance.
5. Thissystemofcoolingisparticularlyadvantageouswherethereareextremeclimatic conditions in the arctic or where there is scarcity of water as in deserts.
6. Noriskofdamage fromfrost,suchascrackingofcylinderjacketsorradiatorwatertubes.

Disadvantagesofaircooledengines

-Relativelylargeamountofpowerisusedtodrivethecoolingfan.™
-Enginesgivelowpoweroutput.™
-Coolingfinsundercertainconditionsmayvibrateandamplifythenoise level.
-Coolingisnotuniform.™
-Enginesaresubjectedtohighworkingtemperature.

Watercoolingsystem:

Cooling water jackets are provided around the cylinder, cylinder head, valve seats etc. The water when circulated through the jackets, it absorbs heat of combustion. This hot water will then be cooling in the radiator partially by a fan and partially by the flow developed by the forward motion of the vehicle. The cooled water is again recirculated through the water jackets.

Antifreezemixture

In western countries if the water used in the radiator freezes because of cold climates,thenice formed has more volume and produces cracks in the cylinder blocks, pipes, and radiator. So, to prevent freezing antifreeze mixtures or solutions are added in the cooling water.

Normallyfollowingareusedasantifreezesolutions:

(i) Methyl,ethylandisopropylalcohols
(ii) Asolutionofalcohol andwater
(iii) EthyleneGlycolraisestheboilingtemperaturesubstantiallyandhencemoreheat dissipation
(iv) AsolutionofwaterandEthyleneGlycol
(v) Glycerinalongwithwater,etc.
(vi) Chromatesareusedtopreventsdeposit

Watercoolingsystemmainlyconsists of:
-Radiator
-Thermostatvalve
-Waterpump
-Fan
-WaterJackets
-Antifreezemixtures

Varioustypesofwatercoolingsystemsaregivenas;

(a) Thermo-syphoncooling
(b) Forcedorpumpcooling
(c) Coolingwiththermostaticregulator
(d) Pressurisedwatercoolingsystem
(e) Evaporativecooling

(a) Thermo-syphoncooling:

This system works on the principle that hot water being lighter rises up and the cold water being heavier goes down. In this system the radiator is placed at a higher level than theengine for the easy flow of water towards the engine. Heat is conducted to the water jackets from where it is taken away due to convection by the circulating water. As the water jacket becomes hot, it rises to the top of the radiator. Cold water from the radiator takes the place of

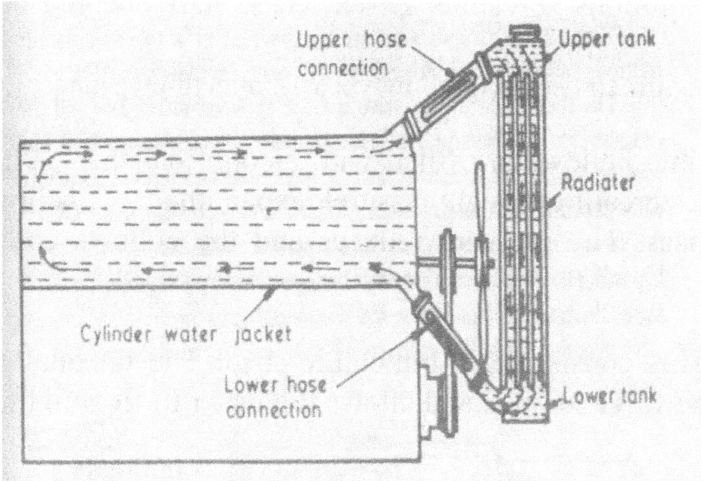

therisinghotwaterandinthiswayacirculationofwaterissetupmthesystem.Thishelpsin keeping the engine at working temperature.

Fig.37.Thermo-syphoncooling

DisadvantagesofThermo-syphonsystem,
-Rateofcirculationistooslow.

-Circulationcommencesonlywhenthereisamarkeddifferenceintemperature.
-Circulationstopsasthelevelofwaterfallsbelowthetopofthedeliverypipeoftheradiator. For these reasons this system has become obsolete and is no more in use.

(b) Forcedorpumpcooling:
-Thissystemisusedinlargenumberofvehicleslikecar,buses,trucksandotherheavy vehicles. Here circulation of water takes place with convection currents help by a pump.

-The water or coolant is circulated through jackets around the parts of the engine to be cooled,andiskeptinmotionbyacentrifugalpump,drivenfromtheenginethroughV-belt.

Limitation
-Coolingisindependentoftemp.
-Engineisovercooled(whenrangeoftemp.=75-90^{0}C)
-Canbeovercomebyusingthermostat

(c) Coolingwiththermostaticregulator:
-Whenever the engine is started from cold, the coolant temperature has to be brought to the desired warm up time to avoid corrosion damage due to condensation of acidsas well as help in easy starting of the engine. This can be done by the use of thermostatic device or thermostat.

-It is a kind of check valve which opens and closes with the effect of temperature. It is fitted in the water outlet of the engine. During the warm-up period, the thermostat is closed and the water pump circulates the water only throughout the cylinder block and cylinder head. When the normal operating temperature is reached, the thermostat valve opens and allows hot water to flow towards the radiator. Standard thermostats are designed to start opening at 70 to 75°C and they fully open at 82°C. High temperature thermostats, with permanent anti-freeze solutions (Prestine, Zerex, etc.), start opening at 80 to 90°C and fully open at 92°C.

(m) Therearethreetypesofthermostats:(i)bellowtype,(ii)bimetallictypeand(iii)waxtype.

*Bellow type valve:*Flexible bellows are filled with alcohol or ether. When the bellows is heated,theliquidvaporises,creatingenoughpressuretoexpandthe bellows.Whentheunitis cooled, the gas condenses. The pressure reduces and the bellows collapse to close the valve.

*Bimetallic type valve:*This consists of a bimetallic strip. The unequal expansion of two metallic strips causes the valve to open and allows the water to flow in the radiator.

Waxtypevalve:

-Canoperatereliablywithinthespecifiedtemperaturerange

(n) Heatistransmittedtowax,whichhashighcoefficientofthermalexpansion

-Uponbeingheated,waxexpandsandtherubberplugpressestheplunger forcingittomove vertically upwards

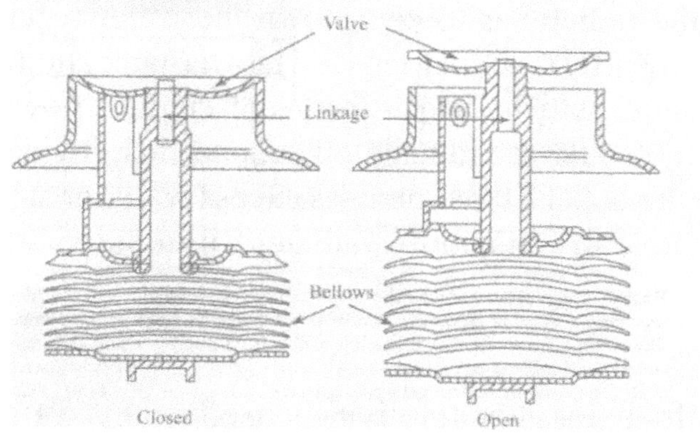

Fig.38.Bellowtypethermostat

(d) Pressurisedwatercooling system:

In the case of the ordinary water-cooling system where the cooling water is subjected to atmospheric pressure, the water boils at 212°F. But, when water is heated in a closed radiator underhighpressure,theboilingtemperatureofwaterincreases.Thehigherwatertemperature gives more efficient engine performance and affords additional protection under high altitude and tropical conditions for long hard driving periods. Therefore, a pressure-type radiator cap is used with the forced circulation cooling system. The cap is fitted on the radiator neck with an air tight seal. The pressure-release valve or safety valve is set to open at a pressurebetween 4 and 13 psi. With this increase in pressure, the boiling temperature ofwater increases to 243°F (at 4 psi boiling tap 225°F and 13 psi boiling temperature 243°F). Any increase in pressure is released by the pressure release valve or safety valve to theatmosphere. On cooling, the vapours will condense and a partial vacuum will be created which will result in the collapse of the hoses and tubes. To overcome this problem the pressure release valve is associated with a vacuum valve which opens the radiator to

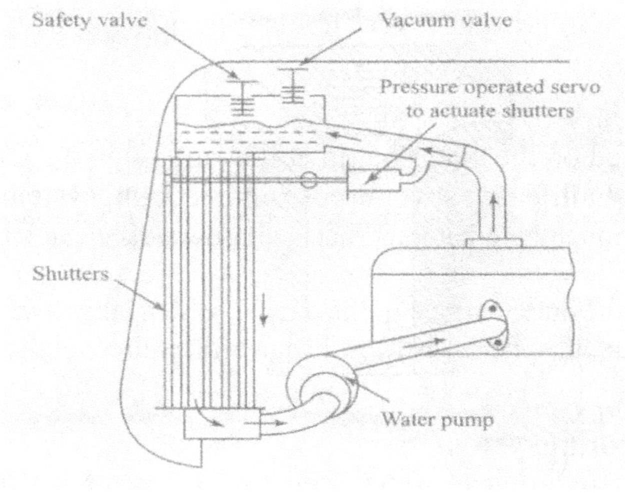

the atmosphere.

Fig.39.Pressurisedsystem

(e) Evaporativecooling system:

-Inthissystem,theenginewillbecooledbecauseoftheevaporationofthewaterinthe cylinder jackets into steams.

-The advantage is being taken from the high latent heat of vaporization of water by allowingit to evaporate in the cylinder jackets. This system is used for cooling of many types of industrial engines

Descriptions of different parts of water cooling systems

Radiator: The purpose of the radiator is to cool down the water received from the engine. The radiator consists of three main parts: (i) upper tank, (ii) lower tank and (iii) tubes.
Hot water from the upper tank, which comes from the engine, flows downwards through the tubes.Theheatcontainedinthehotwaterisconductedtothecopperfinsprovidedaroundthe tubes. An overflow pipe, connected to the upper1 tank, permits excess water or steam to escape.
There are three types of radiators: (i) gilled tube radiator, (ii) tubular radiator and (iii) honey comb or cellular radiator

Gilledtuberadiator:
Thisisperhapstheoldesttypeofradiator,althoughitisstillinuse.In this,waterflowsinside the tubes. Each tube has a large number of annular rings or fins pressed firmly over itsoutside surface.
Tubular radiator: The only difference between a gilled tubes radiator and a tubular one isthatinthiscasethereareno separatefinsforindividualtubes.Theradiatorverticaltubespass through thin fine copper sheets which run horizontally.
Honey comb or cellular radiator: The cellular radiator consists of a large number of individual air cells which are surrounded by water. In this, the clogging ofany passage affects only a small part of the cooling surface. However, in the tubular radiator, if one tube becomes clogged, the cooling effect of the entire tube is lost.

Water Pump:
This is a centrifugal type pump. It is centrally mounted at the front of the cylinder block and is usually driven by means of a belt. This type of pump consists of the following parts: (i) bodyorcasing,(ii)impeller(rotor),(iii)shaft,(iv)bearings,orbush,(v)waterpumpsealand (vi)pulley.
The bottom of the radiator is connected to the suction side of the pump. The power is transmitted to the pump spindle from a pulley mounted at the end of the crankshaft. Seals of various designs are incorporated in the pump to prevent loss of coolant from the system.

Fan:
The fan is generally mounted on the water pump pulley, although on some engines it is attacheddirectlytothecrankshaft.Itservestwopurposesinthecoolingsystemofaengine.
(i) Itdrawsatmosphericairthroughtheradiatorandthusincreasestheefficiencyofthe radiator in cooling hot water.
(ii) It throws fresh air over the outer surface of the engine, which takes away the heat conductedbytheenginepartsandthusincreasestheefficiencyoftheentirecoolingsystem.

***Advantagesofwatercoolingsystem**
-Becauseofevencoolingofcylinderbarrelandhead(duetojacketing)makesitpossibleto reduce the cylinder head and valve seat temperatures.
-Thevolumetricefficiencyofwatercooledenginesishigherthanthatofair-cooledengines.
-Compactdesignofengineswithappreciablysmallerfrontalareaispossible.
-Incaseofwatercooledengines,installationisnotnecessarilyatthefrontofthemobile vehicles, aircraft etc. as the cooling system can be conveniently located.

***Disadvantagesofwatercoolingsystem**
-Thesystemrequiresmoremaintenance.
-Theengineperformancebecomessensitivetoclimaticconditions.™
-Thepowerabsorbedbythepumpisconsiderableandaffectsthepoweroutputofthe engine.™
-Intheeventoffailureofthecoolingsystemseriousdamagemaybecausedtotheengine.

LUBRICATION SYSTEM

IC engine is made of many moving parts. Due to continuous movement of two metallic surfaces over each other, there is wearing moving parts, generation of heat and loss of power in the engine. Hence, lubrication of moving parts is essential to prevent all these harmful effects.

Inenginethefrictionallossesisattributedduetothefollowingmechanicallosses;

(i) Directfrictional losses:
-powerabsorbedduetotherelativemotionofdifferentbearingsurfacessuchaspistonrings, main bearings, cam shaft bearings etc.

(ii) Pumpingloss:
-netpowerspentbythepistononthegasduringintakeandexhaust stroke
-moreincaseoffourstrokeenginecomparedtotwostrokeengine

(iii) Powerlosstodrivecomponentstochargeand scavenge:
-Infourstrokesuperchargedengine,compressorusedtoprovidehighpressureairwhichis mechanically driven by the engine. This is counted as negative frictional loss.
-Intwo-strokeengine scavengingpumpisusedwhichisalsodrivenbythe engine

(iv) Powerlosstodrivetheauxiliaries:
-Somepowerisusedtodriveauxiliariessuchaswaterpump,lubricatingoilpump,fuel pump, cooling fan, generator etc.

Functionof lubrication:
Lubrication produces the following effects: (a) Reducing friction effect (b) Cooling effect (c) Sealing effect and (d) Cleaning effect.

(a) Reducing frictional effect: The primary purpose of the lubrication is to reduce friction and wear between two rubbing surfaces. Two rubbing surfaces always produce friction. The continuous friction produce heat which causes wearing of parts and loss of power. In order to avoid friction, the contactoftwo sliding surfaces must be reduced as far as possible. This can be done by proper lubrication only. Lubrication forms an oil film between two moving surfaces. Lubrication also reduces noise produced by the movement of two metal surfaces over each other.
(b) Cooling effect: The heat, generated by piston, cylinder, and bearings is removed by lubrication to a great extent. Lubrication creates cooling effect on the engine parts.
(c) Sealing effect: The lubricant enters into the gap between the cylinder liner, piston and piston rings. Thus, it prevents leakage of gases from the engine cylinder.
(d) Cleaning effect: Lubrication keeps the engine clean by removing dirt or carbon from inside of the engine along with the oil.

Lubrication theory: There are two theories in existence regarding the application of lubricants on a surface: (i) Fluid film theory and (ii) Boundary layer theory.

(i) Fluid film theory: According to this theory, the lubricant is, supposed to act like mass of globules, rolling in between two surfaces. It produces a rolling effect, which reducesfriction.

(ii) Boundary layer theory: According to this theory, the lubricant is soaked in rubbing surfaces and forms oily surface over it. Thus the sliding surfaces are kept apart from each other, thereby reducing friction.

PropertiesofLubricant:

1. *Viscosity:*Viscosity is a measure of the resistance to flow or the internal friction of the lubricant.
-usually measured by Saybolt universal seconds (SUS) and Redwood viscometer. Also it is expressed with centistoke (unit of kinematic viscosity) and centipoise (unit of absolute viscosity)
-expressedintwotemperaturei.e.-18°C(0°F)and99°C(210°F)

2. *ViscosityIndex:*Itisusedtogradelubricants.Viscosityisinverselyproportionaltotemp.
-If temp. increases, the viscosity of the lubricant decreases and if temp. decreases, the viscosity of the lubricant increases.
-Thevariationofviscosityofoilwithchangesintemperatureismeasuredbyviscosityindex
-oil to measure is compared with 2 reference oil having same viscosity at 99°C. one is paraffinic base oil index of zero and another naphthenic base oil index of 100
-high viscosity index number indicates relatively smaller change in viscosity of the oil with temperature.
-low viscos oil is recommended for automobile engines in winter than summer. The viscosity of a lubricant should be just sufficient to ensure lubrication. If it is more than this value, power loss will be higher due to increased oil resistance.
-VIimproverareaddedtoimproveviscosityindex

3. *Oiliness:*Itis the propertyofa lubricatingoil to spread &attach itselffirmlyto the bearing surfaces as well as provide lubricity. Generally, the oiliness of the lubricating oil should be high particularly when it is to be used for mating surfaces subjected to a high intensity of pressure and smaller clearance portion to avoid the squeezing out of the oil. Such a way that the metal is protected bya thin layer of the oil and the wear is also considerablyreduced. It is measured by co-efficient of friction at extreme operating condition.

4. *Flash Point:*Flash point ofoil is the min. temp. at which the vapours of lubricatingoil will flash when a small flame is passed across its surface. It is of two type open flash point and closed flash point. The flash point of the lubricating oil must be higher thanthe temp. likelyto be developed in the bearings in order to avoid the possibility of fire hazards.

5. *FirePoint:*Ifthelubricatingoilisfurtherheatedaftertheflashpointhasbeenreached,the lowest temp. at which the oil will burn continuously for 5 seconds is called fire point.
-usually11°Chigherthanopenflashpointandvariesfrom190°Cto290°Cforthelubricants used for IC engines
-Thefirepointofalubricantalsomustbehighsothattheoildoesnotburninservice.

6. *Cloud Point:*It is the temp. at which the lubricating oil changes its state from liquid to solid. Its temp. must high for the low temp. operability of the lubricating oil duringwinter.

7. *Pour Point:*It is the lowest temp. at which the lubricating oil will not flow or totally form wax or solidify. This property must be considered because of its effect on starting an engine in cold weather. Oil derived from paraffinic crudes tends to have higher pour points than those derived from naphthenic crudes. The pour points can be lower by the addition of pour pointdepressantusuallyapolymerisedphenolorester.Pourpointmustbeatleast15°Flower than the operating temperature to ensure maximum circulation.

8. *Corrosiveness:*The present of acid (mineral acid, petroleum acid) is harmful to the metal surfaces. The lubrication oil should notattackchemicallythematerialsoftheengine. Thelubricantshouldnotbecorrosive,butitshouldgiveprotectionagainstcorrosion.New oil has low neutralisation number i.e. it maintains the alkaline and acid solution to make the oil neutral.

9. *Oxidation stability:*Itis resistance to oxidation. Due to oxidation the oil will form deposits on the piston rings and lose its lubricating property. Low temperature operation avoiding the hot-area contact and crankcase ventilation can help in preserving the stability of oil over longer periods. Oxidation inhibitors are used to improve oxidation stability. These are complex compounds of sulphur and phosphorus or amine and phenol derivatives.

10. *Cleanliness:*Lubricating oil must be clean. It should not contain dust and dirt particles as well as water content which promote corrosion.

11. *Carbon residue:*after evaporation of a mass sample of lubricating oil under specific condition may remain as carbonaceous residue. It indicates the deposit characteristics of oil. Paraffinic oil has higher carbon residues than the naphthenic base oil.

***Types of lubricants*:**

Lubricantsareatfollowingthree types.

1. Solid:graphic,micaetc
2. Semisolid:grease
3. Liquid:Lubricantsareobtainedfromanimalfat,vegetablesandminerals.Lubricantsmade of animal fat, does not stand much heat. It becomes waxy and gummy which is not very suitable for machines. Vegetable lubricants are obtained from seeds, fruits and plants. Cottonseed oil, olive oil, linseed oil and castor oil are used as lubricant in small machines. Mineral lubricants are most popular for engines and machines. It is obtained from crude petroleum found in nature. Petroleum lubricants are less expensive and suitable for internal combustion engines.

-Graphiteisoftenmixedwithoiltolubricateautomobilespring.Graphiteisalsousedasa cylinder lubricant.

-Greaseisusedforchassislubrication.

Gradeoflubricants: GenerallylubricatingoilsaregradedbySAE(societyofautomotive engineers)methodbyassignanumbertooilwhoseviscosityatgiventemperaturesfallsin certain range.

-Twotemperatures-18°C(0°F)and99°C(210°F)areusedtoassignthe number

ex-*Singlegradetype:*(a)SAE5w,10wand20wgradesareviscosityat-18°C(0°F)and for winteruse.

(b)SAE20,30,40and50gradeslubricatingoilareviscosityat99°C(210°F)andforsummer use.

*Multi-gradetype:*ex-SAE20W/50oilhasviscosityequaltothatatSAE20Wat -18°Cand viscosity equal to that at SAE 50 at 99°C

SAE grades of oil are based on viscosity but not quality based. API (American petroleum institute)usedregular,premiumandheavydutytypeoilwhich arebaseduponpropertiesof oilandoperatingconditions.Generallyregulartypeoilisstraightmineraloil,premiumtype containedoxidationinhibitorsandheavytypecontainedoxidationinhibitorsplusdetergent- dispersant additives.

AccordingtoAPI,

Forgasolineengine5serviceratingsoilareused:SA,SB,SC,SDandSE For diesel

engines 4 service ratings are use: CA, CB, CC and CD
WhereSandCstandsfor SIandCIengines
RatingAisforlight-dutyservice,theseverityofserviceincreasing towardsratingDandE which is severe duty.

Lubricationsystem: variouslubricationsystemusedforICenginesare,

(a) Mistlubrication system
(b) Wetsumplubrication system
(c) Drysumplubricationsystem

(a) Mistlubrication system:

-Usedwherecrankcaselubricationisnotsuitable

(o) Generallyadoptedintwostrokepetrolenginelinescooterandmotorcycle. Itisthesimplest form of lubricating system.

(p) Itisthesimplestformoflubricatingsystem. Itdoesnotconsistofanyseparatepartlikeoil pump for the purpose of lubrication.

(q) In this system the lubricating oil is mixed into the fuel (petrol) while filling in the petrol tank of the vehicle in a specified ratio (ratio of fuel and lubricating oil is from 12:1 to 50:10as per manufacturers specifications or recommendations.

(r) When the fuel goes into the crank chamber during the engine operation, the oil particles go deep into the bearing surfaces due to gravity and lubricate then. The piston rings, cylinder walls, piston pin etc. are lubricated in the same way.

-If the engine is allowed to remain unused for a considerable time, the lubricating oil separates oil from petrol & leads to clogging (blocking) of passages in the carburettor, results in the engine starting trouble. This is the main disadvantage of this system.
-Itcausesheavyexhaustsmokeduetoburningoflubricatingoilpartiallyor fully
-Increasedepositsonpistoncrownandexhaustportswhichaffectengineefficiency
-Corrosionofbearingsurfacesduetoacidsformation
-thoroughmixingcanfetcheffectivelubrication
-Engine suffers insufficient lubrication during closed throttle i.e. vehicle moving down thehill.

(b) Wetsumplubricationsystem:

Bottomofthecrankcasecontainsoilpanorsumpfromwhichthelubricatingoilispumpedto various engine components by a pump. After lubrication, oil flows back to the sump by gravity. Three types of wet sump lubrication system,

(i)
(ii) Splashandpressuresystem
(iii) Pressurefeedsystem

(i) Splashsystem:

-Inthissystem oflubrication the lubricatingoil isstored in an oil sump. Ascoop or dipper is madeinthelowerpartoftheconnectingrod.Whentheengineruns,thedipperdipsintheoil once in every revolution of the crank shaft, the oil is splashed on the cylinder wall. Due to this action engine walls, piston ring, crank shaft bearings are lubricated.

-Itisusedforlightdutyengine

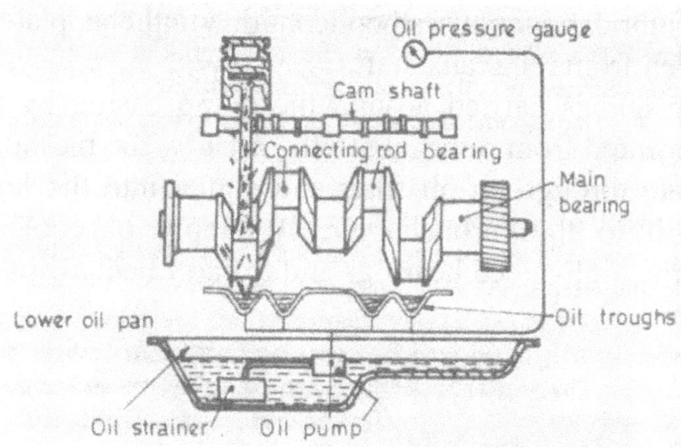

Fig.40.Splashlubricatingsystem

(ii) Splashandpressuresystem:

Lubricating oil is supplied under pressure to main, camshaft bearings and pipeswhich directa stream of oil against the dippers on the big end of connecting rod bearing cup and thus crankpin

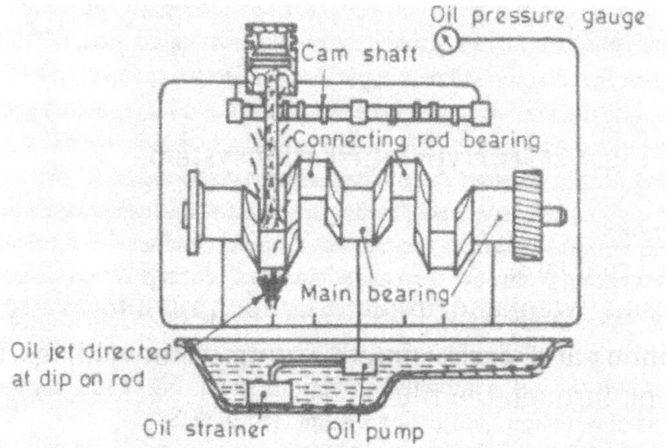

bearings are lubricated by the splash or spray of oil thrown up by the dipper.

Fig.41.Splashandpressurelubricatingsystem

(iii) Pressurefeedsystem:

In this system of lubrication, the engine parts are lubricated under pressure feed. The lubricatingoilisstoredinaseparatetank(incaseofdrysumpsystem)orinthesump(incase of wet sump system), from where an oil pump (gear pump) delivers the oil to the main oil galleryat a pressure of 2-4 kg/cm^2 through an oil filter. The oil from the main gallery goes to main bearing, from where some of it falls back to the sump after lubricating the main bearing and some is splashed to lubricate the cylinder walls and remaining goes through a hole to thecrank pin. From the crank pin the lubricating oil goes to the piston pin through a hole in the connecting rod, where it lubricates the piston rings. For lubricating cam shaft and gearstheoil is led through a separate oil line from the oil gallery. The oil pressure gauge used in the system indicates the oil pressure in the system. Oil filter & strainer in the system clear off the oil from dust, metal particles and other harmful particles.

Fig.42.Pressurelubricatingsystem

Gear system:

-Usedforamediumsizeddieselengine

-Itisaforced-feedsystemoflubricationandusestheoilcontainedinthebedplateas a reservoir. A gear type oil pump is driven from the crankshaft.

-The oil enters the pump and is carried around the pump casing by the gear teeth. It is then discharged. The oil is prevented from returning to the inlet by the meshing of the gear teeth. Oil is pumped from the bed plate through an oil filter and cooler into the lubricating oil manifold. A separate pipe supplies oil to the turbocharger. A supply of cooled oil is critical for the turbocharger to lubricate the high-speed bearings and to carry heat away from the rotor.

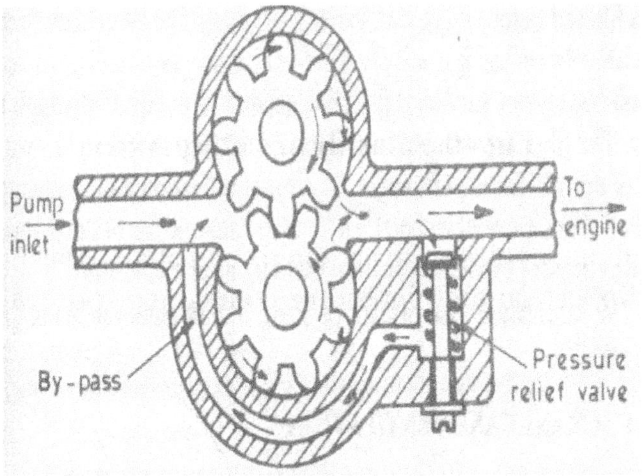

Fig.43.Geartypelubricationsystem

(c) Drysumplubricationsystem:

-Supplyofoil iscarriedinexternal tank

-Oilpumpdrawsoilfromthesupplytankandcirculatesitunderpressuretovariousbearings of the engine

-Oildrippingfromthecylindersandbearingsintothe sumpisremovedbyascavengingpump and again return to supply tank through the filter

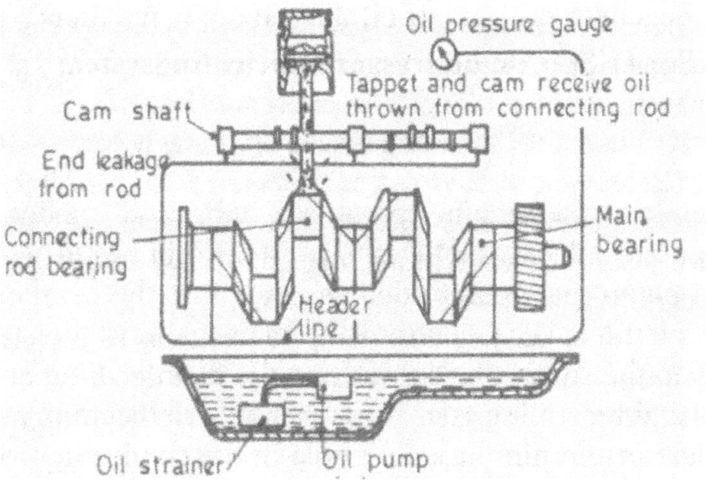

-Thecapacityofscavengingpumpisgreaterthantheoilpump

-Separateoilcoolertoremoveheatfromoilisusedwhichiseithercooledbyairorwater

ModuleIV

Air Compressorss

Functionofnozzle-application-types,Flowofsteamthroughnozzles-Isentropicflow-velocityofnozzleat exit - ideal and actual expansion in nozzle - condition for maximum discharge, critical pressure ratio - super saturated flow, its effect. Types, Impulse and Reaction principles, Compoundings, Velocity diagrams for impulse and reaction blades, Work done on turbine blades and efficiency of components, Speed regulations, governors.

4.1 Introduction

Compressionofairandvapourplaysanimportantroleinengineeringfields.Compressionof air is mostly used since it is easy to transmit air compared with vapour.

4.2 Uses of compressed air:

Theapplicationsofcompressedairarelistedbelow:

1) Itisusedingasturbinesandpropulsionunits.
2) Itisusedinstrikingtypepneumatictoolsforconcretebreaking,clayorrockdrilling, chipping, caulking, riveting etc.
3) Itisusedinrotarytypepneumatictoolsfordrilling,grinding,hammeringetc.
4) Pneumaticliftsandelevatorsworkbycompressedair.
5) Itisusedforcleaningpurposes
6) Itisusedasanatomiserinpaintsprayandinsecticidesspray guns.
7) Piledrivers,extractors,concretevibratorsrequirecompressedair.
8) Air-operatedbrakesareusedinrailwaysandheavyvehiclessuchasbusesandlorries.
9) Sandblastingoperationforcleaningofironcastingsneedscompressedair.
10) Itisused forblastfurnacesandair-operatedchucks.
11) CompressedairisusedforstartingI.C.enginesandalsosupercharging them.

4.3 Working Principles of a Compressor

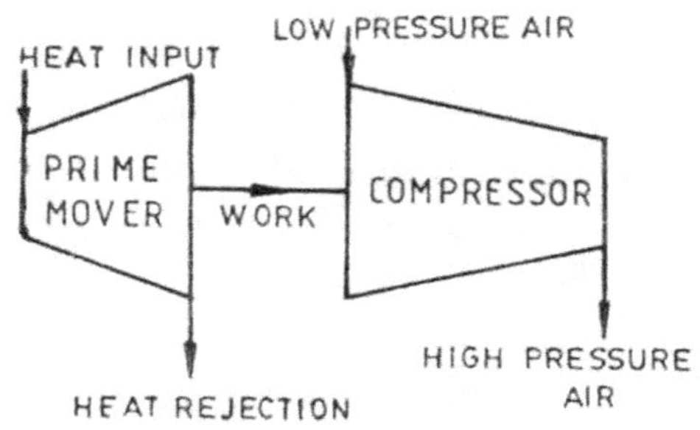

Fig:4.1Air Compressor

A line diagram of a compressor unit is shown in fig:4.1. The compression process requires work input. Hence a compressor is driven by a prime mover. Generally, an electric motor is used as prime mover. Air from atmosphere enters into the compressor It is compressed to a high pressure. Then, this high pressure air is delivered to a storage vessel (reservoir). From the reservoir, it can be conveyed to the desired place through pipe lines.

Some of the energy supplied by the prime mover is absorbed in work done against friction. Some portion of energy is lost due to radiation and coolant. The rest of the energy is maintained within the high pressure air delivered.

4.3 Classificationofcompressors:

Aircompressorsmaybeclassifiedasfollows:

Accordingtodesignandprincipleofoperation:

(a) Reciprocatingcompressorsinwhichapistonreciprocatesinsidethe cylinder.

(b) Rotarycompressorsinwhicharotorisrotated.

According to number of stages:

(a) Singlestagecompressorsinwhichcompressionofairtakesplaceinonecylinder only.

(b) Multistagecompressorsinwhichcompressionofairtakesplaceinmorethanone cylinder.

Accordingtopressurelimit:

(a) Lowpressurecompressorsinwhichthefinaldeliverypressureislessthan10bar,

(b) Mediumpressurecompressorinwhichthefinaldeliverypressureis10barto80bar and

(c) Highpressurecompressorsinwhichthefinaldeliverypressureis80to100bar.

Accordingtocapacity:

(a) Lowcapacitycompressor(delivers0.15m^3 /$s_3$ofcompressedair),

(c) Mediumcapacitycompressor(delivers5m /so3fcompressedair)and Highcapacitycompressor(deliversmorethan5m/sofcompressedair).

According to method of cooling:

(a) Aircooledcompressor (Airisthecoolingmedium) *and*

(b) Watercooledcompressor(Wateristhecoolingmedium).

According to the nature of installation:

(a) Portablecompressors(canbemovedfromoneplacetoanother).

(b) Semi-fixedcompressorsand

(c) Fixedcompressors(Theyarepermanentlyinstalledinone place).

According to applications:

(a) Rockdrillcompressors(usedfordrillingrocks),

(b) Quarryingcompressors(usedinquarries),

(c) Sandblastingcompressors(usedforcleaningofcastiron)and

(d) Spraypaintingcompressors(usedforspray painting). According to number of air cylinders

(a) Simplex-containsoneaircylinder

(b) Duplex-containstwoaircylinders

(c) Triplex-containsthreeair cylinders

4.4.1 Reciprocatingcompressors maybeclassifiedasfollows:

(a) Singleactingcompressorsinwhich suction,compressionanddeliveryofair(orgas) take place on one side of the piston.

(b) Doubleactingcompressorsinwhichsuction,compressionanddeliveryofair(orgas) take place on both sides of the piston.

4.4 Singlestagereciprocatingaircompressor:

In a single stage compressor, the compression of air (or gas) takes place in a single cylinder.Aschematicdiagramofasinglestage,singleactingcompressorisshowninfig:4.2.

Construction: Itconsistsofapistonwhich reciprocatesinsideacylinder.Thepistonis connected to the crankshaft by means of a connecting rod and a crank. Thus, the rotary movement of the crankshaft is converted into the reciprocating motion of the piston. Inlet and outlet valves (suction

anddeliveryvalves)areprovidedatthe top of the cylinder.

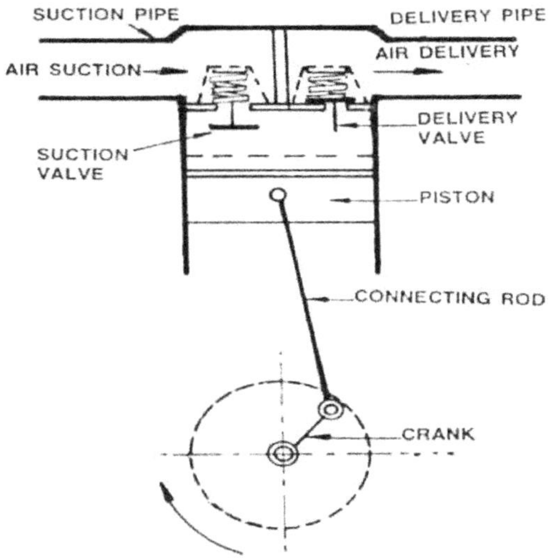

Fig :4.2 Single stage reciprocating AirCompressor

Working: When the piston moves down, the pressure inside the cylinder is reduced. When the cylinder pressure is reduced below atmospheric pressure, the inlet valve opens. Atmospheric air is drawn into the cylinder till the piston reaches the bottom dead centre. The delivery valve remains closed during this period. When the piston moves up, the pressure inside the cylinder increases. The inlet valve is closed, since the pressure inside the cylinder is above atmospheric. The pressure of air inside the cylinder is increased steadily. The outlet valve is then opened and the high pressure air is delivered through the outlet valve into the delivery pipe line.

At the top dead centre of the piston, a small volume of high pressure air is left in the clearance space. When the pistonmoves down again, this air is expanded and pressure reduces, Again the inlet valve opens and thus the cycle is repeated.

Disadvantages

1. Handlingofhighpressure airresultsinleakagethroughthepiston.
2. Coolingofthegas *isnot effective.*
3. Requiresastrongercylindertowithstandhighdeliverypressure.

Applications:Itisusedinplaceswheretherequiredpressureratioissmall.

4.5 Compressionprocesses:

Theairmaybecompressedbythe followingprocesses.

(a) Isentropicoradiabaticcompression,

(b) Polytropiccompression and

(c) Isothermalcompression

(a) Isentropic(or)adiabaticcompression:

Ininternalcombustionengines,theair(orairfuelmixture)iscompressedisentropically. By isentropic compression, maximum available energy in the gas is obtained.

(b) Polytropic compression:

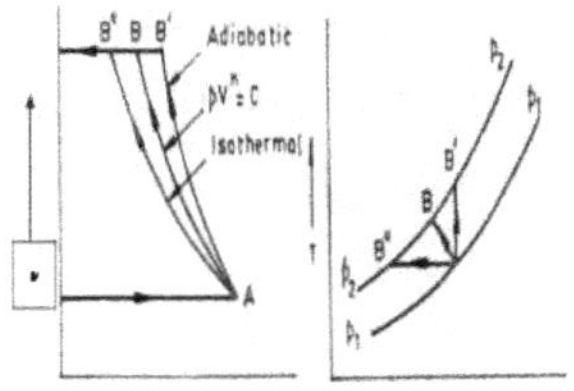

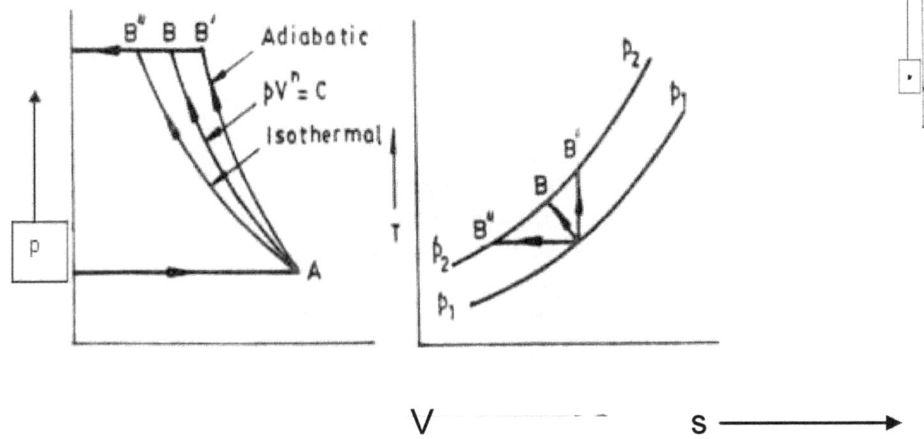

Fig:4.3Compressionprocesses
A-B":Isothermal;A-B:Polytropic;A-B':Isentropic

ThecompressionfollowsthelawpV^n=Constant.Thistypeofcompressionmaybeusedin Bell-Colemancycleofrefrigeration.

(c) Isothermalcompression:

When compressed air (or gas) is stored in a tank, it loses its heat to the surroundings. It attains the temperature of surroundings after some time. Hence, the overall effect of this compression process is to increase the pressure of the gas keeping the temperature constant. Thus isothermal compression is suitable if the compressed air (or gas) is to be stored.

4.6 Powerrequiredfordrivingthecompressor:

Thefollowingassumptionsaremadeinderivingthepowerrequiredtodrivethecompressor.

1. Thereisnopressuredropthroughsuctionanddeliveryvalves.
2. Completecompressionprocesstakesplaceinonecylinder.
3. Thereisnoclearancevolumeinthecompressor cylinder.
4. Pressureinthesuctionlineremainsconstant.Similarly,pressureinthedeliveryline remains constant.
5. Theworkingfluidbehavesasaperfectgas.
6. Thereisnofrictionallosses.

Thecyclecanbeanalysedforthethreedifferentcaseofcompression.Workrequiredcanbe obtained from the p - V diagram.

Let,

p_1=Pressureoftheair(kN/m^2),beforecompression

V_1=Volumeof the air(m^3),beforecompression

T_1=Temperatureoftheair(K),beforecompression

p_2,V_2andT_2bethecorrespondingvaluesaftercompression.

m-Massofairinducedordeliveredbythecycle(kg).

N-Speed inRPM.

4.7.1 Polytropic Compression

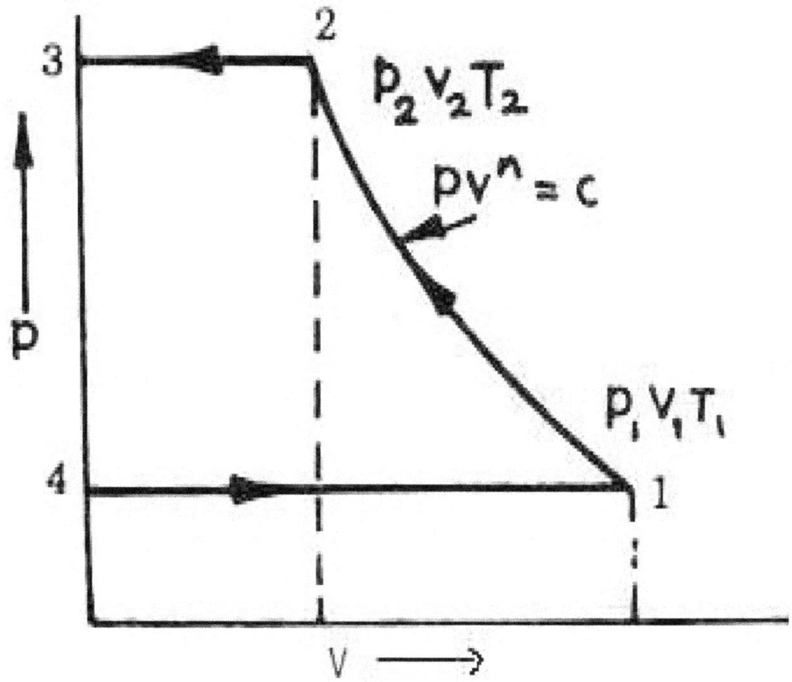

Fig:4.4Polytropiccompression(CompressionfollowspV^n=Constant)

Let n= Index of polytropic compression

Net work done on air/cycle is given by

W = Area 1-2-3-4-1

= Work done during compression (1-2) + Work done during air delivery (2-3) - Work done during suction (4-1).

$$W = \frac{p_2v_2 - p_1v_1}{n-1} + p_2v_2 - p_1v_1$$

$$W = \frac{p_2v_2 - p_1 + (n-1)p_2v_2 - (n-1)p_1v_1}{n-1}$$

$$= \frac{np_2v_2 - np_1v_1}{n-1} = \left(\frac{n}{n-1}\right) p_2v_2 - p_1v_1$$

We know that, $p_1V_1 = m\,RT_1$ & $p_2V_2 = m\,RT_2$

Therefore, $\mathbf{W = \frac{n}{n-1}\, m\, R\, (T_2 - T_1)}$

$$\mathbf{W = \frac{n}{n-1}\, m\, R\, T_1 \left[\frac{T_2}{T_1} - 1\right]}$$

For polytropic process, $\frac{T_2}{T_1} = \left(\frac{p_2}{p_1}\right)^{\frac{n-1}{n}}$

Therefore, $W = \frac{n}{n-1}\, m\, R\, T_1 \left[\left(\frac{p_2}{p_1}\right)^{\frac{n-1}{n}} - 1\right]$ kJ/cycle

$$W = \frac{n}{n-1}\, p_1\, V_1 \left[\left(\frac{p_2}{p_1}\right)^{\frac{n-1}{n}} - 1\right] \text{ kJ/cycle}$$

Indicatedpower(or)Powerrequired,P=WxN,kWforsingleacting reciprocating compressor;

=Wx2N,kWfordoubleactingreciprocatingcompressor.

4.7.2 Isentropiccompression

Compression follows, pV^{γ}= Constant

Let γ = Index of isentropic compression

Net work done on air/cycle is given by

W = Area 1-2-3-4-1

= Work done during compression (1-2) + Work done during air delivery (2-3) - Wc done during suction (4-1).

$$W = \frac{p_2v_2 - p_1v_1}{\gamma-1} + p_2v_2 - p_1v_1$$

$$W = \frac{p_2v_2 - p_1 + (\gamma-1)p_2v_2 - (\gamma-1)p_1v_1}{\gamma-1}$$

$$= \frac{\gamma p_2v_2 - \gamma p_1v_1}{\gamma-1} = \left(\frac{\gamma}{\gamma-1}\right) p_2v_2 - p_1v_1$$

We know that, $p_1V_1 = m\,RT_1$ & $p_2V_2 = m\,RT_2$

$$\mathbf{W = \frac{\gamma}{\gamma-1}\, m\, R\, (T_2 - T_1)}$$

$$\mathbf{W = \frac{\gamma}{\gamma-1}\, m\, R\, T_1 \left[\frac{T_2}{T_1} - 1\right]}$$

For isentropic process, $\frac{T_2}{T_1} = \left(\frac{p_2}{p_1}\right)^{\frac{\gamma-1}{\gamma}}$

Therefore, $W = \frac{\gamma}{\gamma-1}\, m\, R\, T_1 \left[\left(\frac{p_2}{p_1}\right)^{\frac{\gamma-1}{\gamma}} - 1\right]$ kJ/cycle

$$W = \frac{\gamma}{\gamma-1}\, p_1\, V_1 \left[\left(\frac{p_2}{p_1}\right)^{\frac{\gamma-1}{\gamma}} - 1\right] \text{ kJ/cycle}$$

4.7.3 Isothermal Compression

Compression follows, pV= Constant

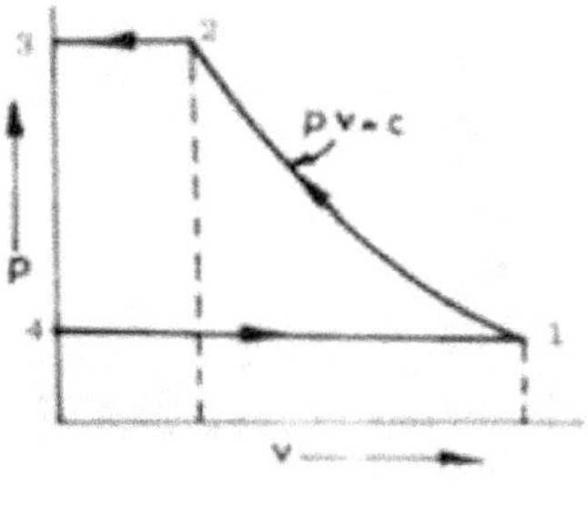

Fig: 4.5 Isothermal Compression

Isothermal Work input, W = Area 1-2-3-4-1 = area under 1-2 + area under 2-3 - area under 4-1

$$W = p_1V_1\, ln\left(\frac{V_1}{V_2}\right) + p_2V_2 - p_1V_1$$

But $p_1V_1 = p_2V_2$

$$W = p_1V_1 \ln\left(\frac{V_1}{V_2}\right) \quad \text{and} \quad \frac{V_1}{V_2} = \frac{p_2}{p_1}$$

Therefore, $W = p_2V_2 \ln\left(\frac{p_2}{p_1}\right)$ kJ/cycle

4.8 Isothermal efficiency: Isothermal efficiency is defined as the ratio of isothermal work input to the actual work input. This is used for comparing the compressors.

$$\text{Isothermal efficiency, } \eta_{iso} = \frac{\text{Isothermal work input}}{\text{Actual work output}}$$

4.9 Adiabatic efficiency: Adiabatic efficiency is defined as the ratio of adiabatic work input to the actual work input. This is used for comparing the compressors.

$$\text{Adiabatic efficiency, } \eta_{adia} = \frac{\text{Adiabatic work input}}{\text{Actual work output}}$$

4.10 Mechanical efficiency:

The compressor is driven by a prime mover. The power input to the compressor is the shaft power (brake power) of the prime mover. This is also known as brake power of the compressor.

Mechanical efficiency is defined as the ratio of indicated power of the compressor to the power input to the compressor.

$$\eta_m = \frac{\text{Indicated power of compressor}}{\text{Power input}}$$

Indicated Power, IP $= \frac{p_m l\, a N k}{60}$,

where, p_m = mean effective pressure, kN/m²

l = length of stroke of piston, m

a = area of cross section of cylinder, m²

N= crank speed in rpm, and

K =number of cylinders

4.11 Clearanceandclearancevolume:

Whenthepistonreachestopdeadcentre(TDC)inthecylinder,thereisadeadspace between piston top and the cylinder head. This space is known as clearance space and the volumeoccupiedbythisspaceisknownasclearancevolume,V_C.

Theclearancevolumeisexpressedaspercentageofpistondisplacement.Itsvalueranges from 5% - 10% of swept volume or stroke volume (V_S).The p - V diagram for a single stage compressor,consideringclearancevolumeisshowninfig..Attheendofdeliveryofhigh pressure air (at point 3), a small amount of high pressure air at p_2 remains in the clearance space. This high pressure air which remains at the clearance space when the piston is at TDC

is known as remnant air. It is expanded polytropically till atmospheric pressure (p_4=p_1) is reached. Theinlet valve is opened and the fresh air is sucked into thecylinder. The suction of airtakesplacefortherestofstroke(uptopoint1).Thevolumeofairsuckedisknownas effective suction volume (V_1 - V_4). At point 1, the air is compressed polytropically till the delivery pressure(p_2) isreached.Thenthedelivery valveisopenedandhighpressureair isdischargedintothereceiver.Thedeliveryofaircontinuestillthepistonreachesitstopdead centre, then the cycle is repeated.

4.11.1 Effect ofclearance volume:

Thefollowingaretheeffectsofclearance space.

1. Suctionvolume(volumeofairsucked)is reduced.
2. Massofairisreduced.
3. Ifclearancevolumeincreases,heavycompressionis required.
4. Heavycompressionincreasesmechanicallosses

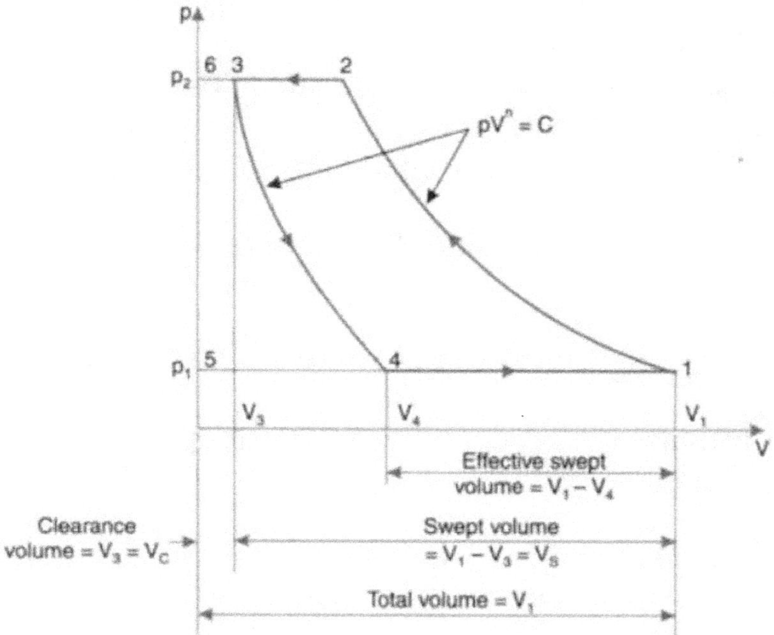

Fig: 4.6 p-V diagram with clearance volume

4.11.3 Work input considering clearance volume:

Assuming the expansion (3-4) and compression (1-2) follow the law $p V^n = C$.
Work input per cycle is given by,
W = Area (1-2-3-6-5-4-1)- Area (3-6-5-4 -3)
W = Workdone during compression - Work done during expansion

$$W = \frac{n}{n-1} p_1 V_1 \left[\left(\frac{p_2}{p_1}\right)^{\frac{n-1}{n}} - 1\right] - \frac{n}{n-1} p_4 V_4 \left[\left(\frac{p_3}{p_4}\right)^{\frac{n-1}{n}} - 1\right]$$

But, $p_3 = p_2$ and $p_4 = p_1$
therefore

$$W = \frac{n}{n-1} p_1 V_1 \left[\left(\frac{p_2}{p_1}\right)^{\frac{n-1}{n}} - 1\right] - \frac{n}{n-1} p_1 V_4 \left[\left(\frac{p_2}{p_1}\right)^{\frac{n-1}{n}} - 1\right]$$

$$W = \frac{n}{n-1} p_1 (V_1 - V_4) \left[\left(\frac{p_2}{p_1} \right)^{\frac{n-1}{n}} - 1 \right] \text{ kJ/cycle}$$

V_1-V_4 is called as effective suction volume.

4.12 Volumetric efficiency:

The clearance volume in a compressor reduces the intake capacity of the cylinder. This lea to a term called volumetric efficiency.

The volumetric efficiency is denned as the volume of free air sucked into the compressor p cycle to the stroke volume of the cylinder.the volume measured at the intake pressure and temperature or at standard atmospheric conditions.(p_s =101.325 kN/m^2 and T_s = 288K)

$$\text{Volumetric efficiency, } \eta_{vol} = \frac{\text{Volume of free air taken in per cycle}}{\text{Stroke volume of the cylinder}}$$

$$= \frac{\text{Effective suction volume}}{\text{Swept volume}} = \frac{(V1 - V4)}{(V1 - V3)} = \frac{V_1 - V_4}{V_s}$$

Clearance ratio: Clearance ratio is defined as, the ratio of clearance volume to swept volum It is denoted by the letter C.

$$\text{Clearance ratio,C} = \frac{\textit{Clearance volume}}{\textit{Swept volume}} = \frac{V_c}{V_s} = \frac{V_c}{V_1 - V_3}$$

$$\text{Pressure ratio, } R_p = \frac{\textit{Delivery pressure}}{\textit{Suction pressure}} = \frac{p_2}{p_1} = \frac{p_3}{p_4}$$

4.12.1 Expression for Volumetric efficiency

Let the compression and expansion follows the law, pV^n =Constant.

$$\text{Clearance ratio, } C = \frac{\textit{Clearance volume}}{\textit{Swept volume}} = \frac{V_c}{V_s} = \frac{V_3}{V_1 - V_3}$$

$V_1 - V_3 = \frac{V_3}{C}$ --------------(1)

$V_1 = \frac{V_3}{C} + V_3$

$V_1 = V_3 \left(\frac{1}{C} + 1\right)$ -------------------- (2)

We know that. Pressure ratio. $R_p = \frac{Delivery\ pressure}{Suction\ pressure} = \frac{p_2}{p_1} = \frac{p_3}{p_4}$

By polytropic expansion process 3-4:

$$\frac{p_3}{p_4} = \left(\frac{V_4}{V_3}\right)^n$$

$$\frac{V_4}{V_3} = \left(\frac{p_3}{p_4}\right)^{1/n} = (R_p)^{\frac{1}{n}}$$

Therefore. $V_4 = V_3(R_p)^{\frac{1}{n}}$ ----------------- (3)

Volumetric efficiency. $\eta_{vol} = \frac{\text{Effective suction volume}}{\text{Swept volume}} = \frac{(V_1 - V_4)}{(V_1 - V_3)}$ ------------ (4)

Using equations 1.2 and 3 in 4.

$$\eta_{vol} = \frac{V_3\left(\frac{1}{C}+1\right) - V_3[R_p]^{1/n}}{\frac{V_3}{C}} = \frac{V_3\left\{\left(\frac{1}{C}+1\right) - [R_p]^{1/n}\right\}}{V_3\left(\frac{1}{C}\right)} = \frac{\left\{\left(\frac{1}{C}+1\right) - [R_p]^{1/n}\right\}}{\left(\frac{1}{C}\right)} = C\left[\left(\frac{1}{C}+1\right) - [R_p]^{1/n}\right]$$

$$\eta_{vol} = 1 + C - C[R_p]^{1/n} = 1 + C - C\left[\frac{p_2}{p_1}\right]^{1/n}$$

4.13 Multi-stageaircompressor:

In a multi stage air compressor, compression of air takes place in more than one cylinder. Multi stageaircompressor is usedin places where high pressure airis required.Fig. shows the general arrangement of a two-stage air compressor. It consists of a low pressure (L.P)cylinder,anintercoolerandahighpressure(H.P)cylinder.Boththepistons(in L.Pand H.Pcylinders)aredrivenbyasingleprimemoverthroughacommonshaft.

Atmospheric air at pressure p_1 taken into the low pressure cylinder is compressed to a high pressure (p_2). This pressure is intermediate between intake pressure (p_1) and delivery pressure p_3). Hence this is known as intermediate pressure.

The air from low pressure cylinder is then passed into an intercooler. In the intercooler, the air is cooled at constant pressure by circulating cold water. The cooled air from the intercooler is then taken into the high pressure cylinder. In the high pressure cylinder, air is further compressed to the final delivery pressure (p_3) and supplied to the air receiver tank.

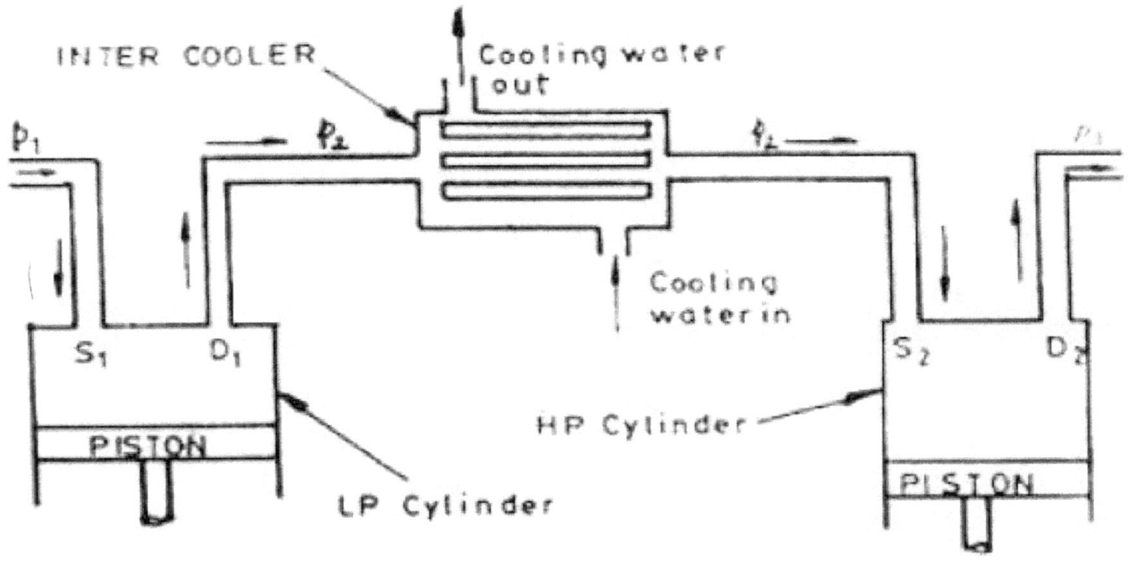

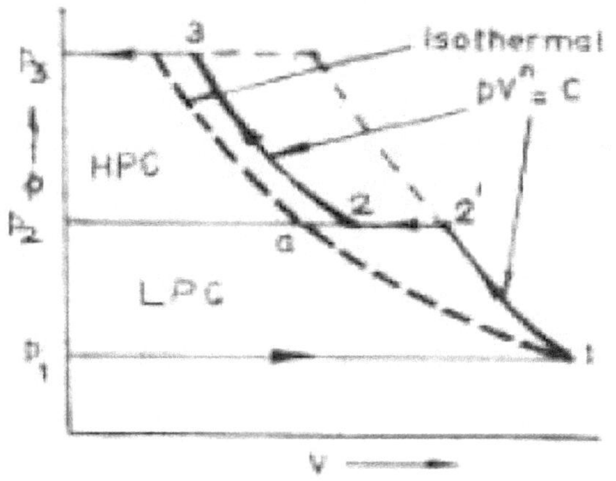

Advantages:

1. Saving in work input: The air iscooledinanintercoolerbeforeenteringthehigh pressure cylinder.Hence less powerisrequiredtodriveamultistagecompressoras compared to a single stage compressor for delivering same quantity of air at the same deliverypressure.

2. Betterbalancing:Whentheairissuckedinonecylinder,thereiscompressionintheother cylinder. This provides more uniform torque. Hence size of the flywheel is reduced.

3. No leakage and better lubrication: The pressure and temperature ranges are kept within desirablelimits.Thisresultsina)Minimumairleakagethroughthepistonofthecylinder and b)effectivelubricationduetolowertemperature.

4. More volumetric efficiency: For small pressure range, effect of expansion of the remnant air (high pressure air in the clearance space) is less. Thus by increasing number of stages, volumetric efficiency is improved.

5. High delivery pressure: The delivery pressure of air is high with reasonable volumetric efficiency.

6. Simple construction of LP cylinder: The maximum pressure in the low pressure cylinder is less. Hence, low pressure cylinder can be made lighter in construction.

7. Cheapermaterials:Loweroperatingtemperaturepermitstheuseofcheapermaterialsfor construction.

1.Morethanonecylinderisrequired.

Disadvantages:

2Anintercoolerisrequired.Thisincreasesinitialcost.Alsospacerequiredismore.
3. Continuousflowofcoolingwater isrequired.
4. Complicatedinconstruction.

4.14 Intercoolers:

An intercooler is a simple heat exchanger. It exchanges the heat of compressed air from the LP compressor to the circulating water before the air enters the HP compressor. It consists of a number of special metal tubes connected to corrosion resistant plates at both ends. The entire nest of tubes is covered by an outer shell

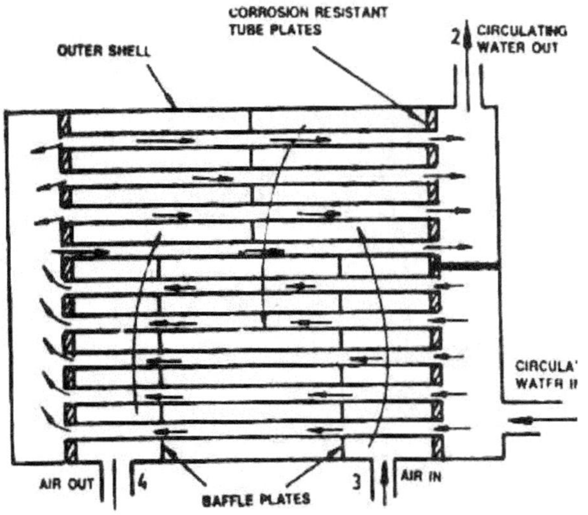

Working: Cold water enters the bottom of the intercooler through water inlet (1) and flows into the bottom tubes. Then they pass through the top tubes and leaves through the water outlet(2)atthetop.AirfromLPcompressorentersthroughtheairinlet(3)oftheintercooler and passes over the tubes. While passing over the tubes, the air is cooled (by the cold water circulated through the tubes). This cold air leaves the intercooler through the air outlet (4).
Baffleplatesareprovidedintheintercoolertochangethedirectionofair.Thisprovidesa better heat transfer from air to the circulating water.

4.15 Workinputrequiredinmultistagecompressor:

The following assumptions are made for calculating the work input in multistage compression.

1. Pressureduringsuctionanddelivery remains constant in each stage.
2. Intercoolingtakesplaceatconstant pressure in each stage.
3. Thecompressionprocessissameforeach stage.
4. ThemassofairhandledbyLPcylinderandHP cylinder

is same.

5. Thereisnoclearancevolumeineach cylinder.

6Thereisnopressuredropbetweenthetwo stages, i.e., exhaust pressure of one stage is

Workrequiredtodrivethemulti-stagecompressorcanbecalculatedfromtheareaofthep -V diagram .

Let,p_1,V_1andT_1betheconditionofairenteringtheLPcylinder.P_2, V_2 and T_2 be the condition of air entering the HP cylinder. $p_3$bethefinaldeliverypressureofair.

Then.

Total work input = Work input for LP compressor + Work input for HP compressor.

$$W = \frac{n}{n-1} p_1 V_1 \left[\left(\frac{p_2}{p_1}\right)^{\frac{n-1}{n}} - 1\right] + \frac{n}{n-1} p_1 V_1 \left[\left(\frac{p_3}{p_2}\right)^{\frac{n-1}{n}} - 1\right] \text{ kJ/cycle}$$

$$W = \frac{n}{n-1} m R T_1 \left[\left(\frac{p_2}{p_1}\right)^{\frac{n-1}{n}} - 1\right] + \frac{n}{n-1} m R T_2 \left[\left(\frac{p_3}{p_2}\right)^{\frac{n-1}{n}} - 1\right] \text{ kJ/cycle}$$

If intercooling is perfect. $T_2 = T_1$, therefore.

$$W = \frac{n}{n-1} m R T_1 \left[\left(\frac{p_2}{p_1}\right)^{\frac{n-1}{n}} - 1\right] + \frac{n}{n-1} m R T_1 \left[\left(\frac{p_3}{p_2}\right)^{\frac{n-1}{n}} - 1\right] \text{ kJ/cycle}$$

$$W = \frac{n}{n-1} m R T_1 \left[\left(\frac{p_2}{p_1}\right)^{\frac{n-1}{n}} + \left(\frac{p_3}{p_2}\right)^{\frac{n-1}{n}} - 2\right] \text{ kJ/cycle}$$

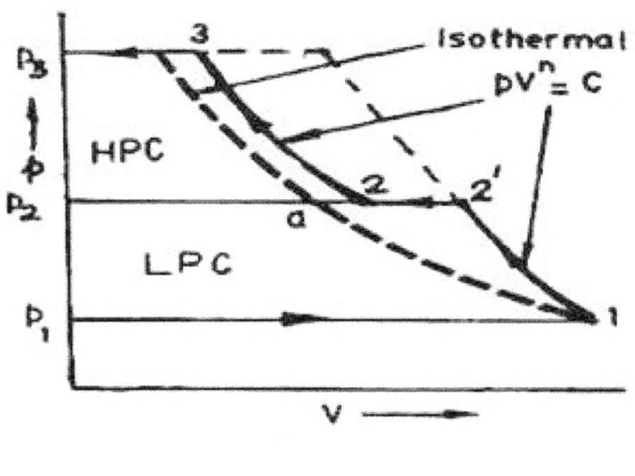

4.16 Condition for maximum efficiency (or)
Condition for minimum work input (or)
To prove that for minimum work input the intermediate pressure of a two-stage compressor with perfect intercooling is the geometric mean of the intake pressure and delivery pressure (or)
To prove $p_2 = \sqrt{p_1 p_3}$

Work input for a two-stage air compressor with perfect intercooling is given by,

$$W = \frac{n}{n-1} p_1 V_1 \left[\left(\frac{p_2}{p_1}\right)^{\frac{n-1}{n}} + \left(\frac{p_3}{p_2}\right)^{\frac{n-1}{n}} - 2 \right] \text{ kJ/cycle}$$

If the initial pressure (p_1) and final pressure (p_3) are fixed, the value of intermediate pressure (p_2) can be determined by differentiating the above equation of work input in terms of p_2 and equating it to zero.

Let, $\frac{n}{n-1} p_1 V_1 = k\ (constant)$ and $\frac{n-1}{n} = a$

then,

$$W = k \left[\left(\frac{p_2}{p_1}\right)^{a} + \left(\frac{p_3}{p_2}\right)^{a} - 2 \right]$$

or

$$W = k(p_2^a\, p_1^{-a} + p_3^a p_2^{-a} - 2) \quad \text{------------------- (1)}$$

Differentiating the above equation (1) with respect to p_2 and equating it to zero,

$$\frac{dW}{dp_2} = k\, a\, p_2^{a-1}\, p_1^{-a} + k\,(-a) p_3^a\, p_2^{-a-1} = 0$$

$$k\, a\, \frac{p_2^a}{p_2 p_1^a} - k\, a\, p_3^a\, \frac{1}{p_2^a\, p_2} = 0$$

or

$$\frac{k\, a\, p_2^a}{p_2 p_1^a} = \frac{k\, a\, p_3^a}{p_2 p_2^a}$$

$$\left(\frac{p_2}{p_1}\right)^a = \left(\frac{p_3}{p_2}\right)^a$$

=> $p_2^2 = p_1 p_3$

or

intermediate pressure, $p_2 = \sqrt{p_1 p_3}$

Thus for maximum efficiency the intermediate pressure is the geometric mean of the initial and final pressures.

For a three stage compressor.

$$W_{min} = \frac{3n}{n-1} p_1 V_1 \left[\left(\frac{p_4}{p_1} \right)^{\frac{n-1}{3n}} - 1 \right]$$

4.17 Minimum work input for multistage compression with perfect intercooling:

Work input for a two-stage compressor with perfect intercooling is given by

$$W = \frac{n}{n-1} p_1 V_1 \left[\left(\frac{p_2}{p_1} \right)^{\frac{n-1}{n}} + \left(\frac{p_3}{p_2} \right)^{\frac{n-1}{n}} - 2 \right] \quad \text{------------------------ (1)}$$

Work input will be minimum if $\frac{p_2}{p_1} = \frac{p_3}{p_2}$ ------------------------------ (2)

$$p_2^2 = p_1 \, p_3$$

Dividing both sides by p_1^2 .

$$\left(\frac{p_2}{p_1} \right)^2 = \frac{p_3}{p_1} \qquad \Rightarrow \frac{p_2}{p_1} = \left(\frac{p_3}{p_1} \right)^{1/2} \quad \text{-------------------------------- (3)}$$

From (2), $\frac{p_3}{p_2} = \frac{p_2}{p_1} = \left(\frac{p_3}{p_1} \right)^{1/2}$ --------------------------------------- (4)

Substituting the equation (4) in equation (1), work input for a two stage compressor,

$$W_{min} = \frac{n}{n-1} p_1 V_1 \left[\left(\frac{p_3}{p_1} \right)^{1/2 \left[\frac{n-1}{n}\right]} + \left(\frac{p_3}{p_1} \right)^{1/2 \left[\frac{n-1}{n}\right]} - 2 \right]$$

$$= \frac{n}{n-1} p_1 V_1 \left[2 \left(\frac{p_3}{p_1} \right)^{\frac{n-1}{2n}} - 2 \right]$$

$$W_{min} = \frac{2n}{n-1} p_1 V_1 \left[\left(\frac{p_3}{p_1} \right)^{\frac{n-1}{2n}} - 1 \right]$$

or

$$W_{min} = \frac{2n}{n-1} mRT_1 \left[\left(\frac{p_3}{p_1} \right)^{\frac{n-1}{2n}} - 1 \right]$$

Minimum work input required for a two stage reciprocating air compressor with perfect intercooling is given by,

$$W_{min} = \frac{2n}{n-1} p_1 V_1 \left[\left(\frac{p_3}{p_1}\right)^{\frac{n-1}{2n}} - 1 \right] kJ$$

But,from equation (4), $\left(\frac{p_3}{p_1}\right)^{1/2} = \frac{p_2}{p_1}$

Therefore,

$$\boldsymbol{W_{min} = \frac{2n}{n-1} p_1 V_1 \left[\left(\frac{p_2}{p_1}\right)^{\frac{n-1}{n}} - 1 \right] kJ}$$

So, for maximum efficiency ie., for minimum work input, the work required for each stage is same.

For maximum efficiency, the following conditions must be satisfied:

1. The air is cooled to the initial temperature between the stages (Perfect cooling between stages).
2. In each stage, the pressure ratio is same.$\left(\frac{p_2}{p_1} = \frac{p_3}{p_2} = \frac{p_4}{p_3} = \cdots\right)$
3. The work input for each stage is same.

MODULEV

Rotarycompressorshavearotortodeveloppressure.Theyareclassifiedas (1)Positivedisplacementcompressorsand(2)Nonpositivedisplacement(Dynamic) compressors

In positive displacement compressors, the air is trapped in between two sets of engagingsurfaces. The pressure rise is obtained by the back flow of air (as in the case of Roots blower) or both by squeezing action and back flow of air (as in the case of vane blower). Example: (1)Roots blower, (2) Vane blower, (3) Screw compressor.
In dynamic compressors, there is a continuous steady flow of air. The air is not positively containedwithincertainboundaries.Energyistransferredfromtherotorofthecompressortothe air. The pressure rise is primarily due to dynamic effects.
Example:(1)Centrifugalcompressor,(2)Axialflowcompressor.

4.18.1Rootsblower: TheRootsblowerisadevelopmentofthegear pump.

Construction: It consists of two lobed rotors placed in separate parallel axis of acasing as shownin fig:4.11.Thetworotorsaredrivenbyapairof gears(whicharedrivenbytheprime mover)andthey revolveinoppositedirections.Thelobesoftherotorareofcycloidshapetoensurecorrectmating.A smallclearanceof0.1mmto0.2 mmisprovidedbetweenthelobeand casing.Thisreducesthewear of moving parts.
Working: When the rotor is driven bythe gear, air istrapped betweenthe lobes and the casing. the trappedairmovesalongthecasinganddischargedintothereceiver.Thereisnoincreaseinpressure since the flow area from entry to exit remains constant. But, when the outlet is opened, there is a backflowofhighpressureairinthereceiver.Thiscreatestheriseinpressureoftheairdelivered. These types of blowers are used in automobiles for supercharging.

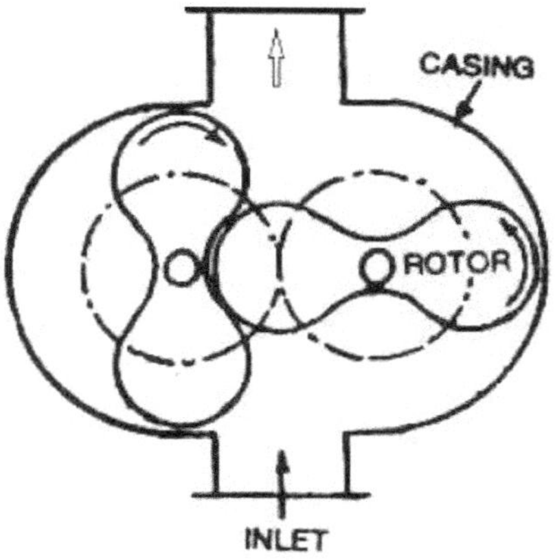

Fig:4.11 Roots blower

Construction: A vane blower consists of (1) a rotor, (2) vanes mounted on the rotor, (3) inlet and outlet ports and (4) casing. The rotor is placed eccentrically in the outer casing. Concentric vanes (usually6to8nos.)aremountedontherotor.Thevanesaremadeoffiberorcarbon. Inletsuction area is greater than outlet delivery area.

Vaneblower

Working: When the rotor is rotated by the prime mover, air is entrapped between two consecutive vanes. This air is gradually compressed due to decreasing volume between the rotor and the outer casing.Thisairisdeliveredtothereceiver.Thispartlycompressedairisfurtherincreasedinpressure due to the back flow of high pressure air from the receiver.

Advantages:1.Verysimpleandcompact,2.Highefficiency3.Higherspeedsarepossible

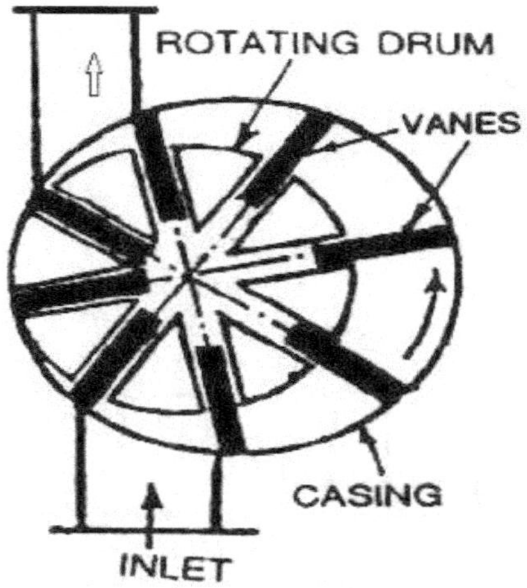

Fig: 4.12 Vane blower

4.18.3 Centrifugalcompressor

Construction: It consists of an impeller, a casing and a diffuser. The impeller consists of a number ofbladesorvanes,ismountedonthecompressorshaft insidethecasing.Theimpellerissurrounded by the casing.

Working: In this compressor air enters axially and leaves radially. When the impeller rotates, air enters axially through the eye of the impeller with a low velocity. This air moves over the impeller vanes. Then, it flows radiallyoutwards fromthe impeller.The velocityand pressure increasesin the impeller. Theair then enters the divergingpassage known as diffuser. In the diffuser, kinetic energy isconvertedintopressureenergyandthepressureoftheairfurtherincreases.Itisshowninfig:4.14. Finally, high pressure air is delivered to the receiver. Generally half of the total pressure rise takes place in

the impeller and the other half in the diffuser.
Applications:Centrifugalcompressorsareusedforlowpressureunitssuchasfor refrigeration, supercharging of internal combustion engines, etc.

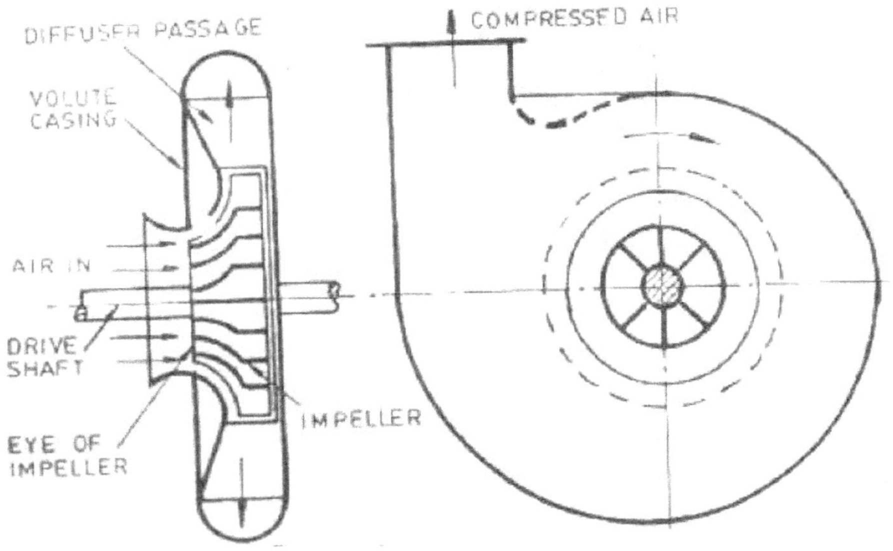

4.18.4 Axialflowcompressor

Inthisaircompressor,airentersandleavesaxially.

Construction: It consists of two sets of blades: Rotorblades and stator blades. The blades are so arranged that the unit consists of adjacent rows of rotor blades and stator blades as shownin fig:4.15. The stator blades are fixed to the casing. The rotor blades are fixed on the rotating drum. The drum is rotated by a prime mover through a driving shaft. Single stage compressor consists of a row of rotor blades followed by a row of stator blades. Compression of air takes place in each pair of blades (one rotor blade and one stator blade). Hence there are many stages of compression in this type of compressor.

Working: When the switch is switched on, the prime mover rotates the drum. Air enters through the compressor inlet and passes through the rotor and stator blades. Whilepassing

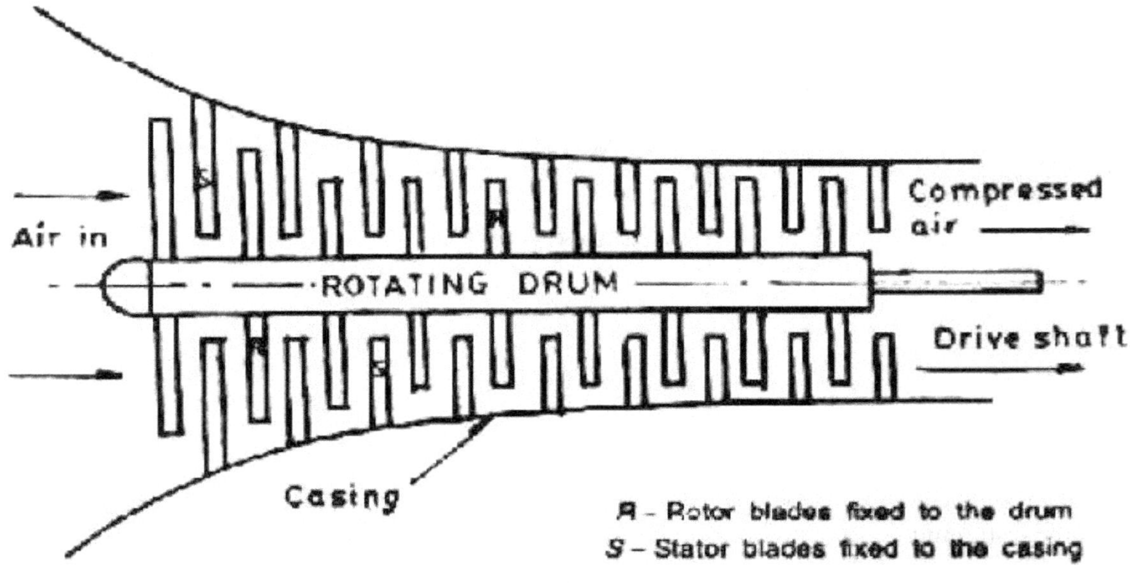

Fig:4.15 Axial flow compressor

Applications:

1. Theyarewidelyusedinhighpressureunitssuchasindustrialandmarinegasturbineplants,
2. Theyaremostsuitableforaircraftwork(Jetpropulsion)sincetheyrequirelessfrontalarea.

QUESTION PAPERS From VTU

Model Question Paper – 1 (CBCS) with effect from 2015-16

USN | | | | | | | | | | | 15ME33

Third Semester B.E. Degree (CBCS) Examination, Dec.2016/Jan.2017

Basic Thermodynamics

Time: 3 hrs.   Max. Marks: 80

Note: 1. Answer any FIVE full questions, choosing one full question from each module.
2. Use of Thermodynamic data handbook and steam table is permitted.

MODULE- 1

1 a. Define the following with examples.
i) Open system ii) Closed system iii) Isolated system (08Marks)

b. State Zeroth law of thermodynamics. The readings t_A and t_B of two Celsius thermometers A and B agree at ice & steam point, but elsewhere are related by the equation t_A=L+Mt_B +Nt_B^2 where L,M,N are constants, when both thermometers are immersed in a system of fluid , A registers 11°C while B registers10°C.Determine the reading on A when B registers 37.4°C (08 Marks)

OR

2. a. Define thermodynamic work. Write similarities & dissimilarities between Heat and Work. (08Marks)

b. A gas initially at 100KPa and 6000cm³. The final volume is 2000cm³. Determine the moving boundary work for each of the following processes.
(i) P is inversely proportional to V (ii) PV^2= constant iii) P is inversely proportional to V (08Marks)

MODULE- 2

3. a. Describe the classic paddle wheel experiment performed by Joule. What conclusion was drawn based on the experimental observations (Joule experiment). (08Marks)

b. A turbine operates under steady flow conditions, receives steam at the following state: Pressure 1.2MPa, temperature 188oC, enthalpy 2785KJ/Kg, velocity 34 m/s & elevation 3m. The steam leaves the turbine at the following state: pressure 20Mpa, enthalpy 2512KJ/kg, velocity 100m/s and elevation 0 m. Heat loss to the surrounding at a rate of 0.29KJ/s. If the steam rate is0.42kg/s.Determine power output from the turbine. (08Marks)

OR

4. a. State and Prove that Kelvin- Planck and Clausis statements of second law of thermodynamics (08Marks)

b. Using a heat engine of thermal efficiency of 30% to drive a refrigerator having a COP of 5, what is the heat received by the heat engine for each MJ of heat removed from the cold body of the refrigerator? (08Marks)

CBCS SCHEME

USN | | | | | | | | | | 18ME33

Third Semester B.E. Degree Examination, Jan./Feb. 2021
Basic Thermodynamics

Time: 3 hrs. Max. Marks: 100

Note: 1. Answer any FIVE full questions, choosing ONE full question from each module.
2. Use of thermodynamic data handbook is permitted.

Important Note : 1. On completing your answers, compulsorily draw diagonal cross lines on the remaining blank pages.
2. Any revealing of identification, appeal to evaluator and /or equations written eg, 42+8 = 50, will be treated as malpractice.

Module-1

1 a. Explain microscopic and macroscopic approaches to thermodynamics. **(06 Marks)**

b. State and explain Zeroth law of thermodynamics. What is diathermal and adiabatic wall? **(06 Marks)**

c. The temperature t on a Celsius thermometer scale is defined in terms of property P by the relation $P = e^{\frac{(t-B)}{A}}$ where A and B are constants. At ice point and steam points the value of P is 1.86 and 6.81 respectively. Find the value of 't' for $P = 2.5$. **(08 Marks)**

OR

2 a. With examples distinguish between:
(i) Intensive and extensive property
(ii) Point and path function
(iii) Thermodynamic equilibrium **(10 Marks)**

b. In 1709m Newton proposed a linear temperature scale where ice point and normal human body temperature are maintained as two fixed points of 0°N ad 12°N respectively. The temperature of human body on the Celsius scale is 36°C. Obtain relation between Newton scale and Celsius scale. **(10 Marks)**

Module-2

3 a. Obtain the expression for displacement adiabatic work. **(06 Marks)**

b. Define heat and work with reference to thermodynamic point of view and also the sign convention of heat and work. **(06 Marks)**

c. A cylinder contains 1 kg of certain fluid at an initial pressure of 20 bar. The fluid is allowed to expand reversibly behind a piston according to law PV^2 = constant until the volume is doubled. The fluid is then cooled reversibly at constant pressure until the piston regains its original position. Heat is then supplied reversibly with the piston firmly locked in position until the pressure rises to its original value of 20 bar. Calculate the net work done by the fluid for an initial volume of $0.05\ m^3$. **(08 Marks)**

OR

4 a. Apply steady flow energy equation to each of the following:
(i) Boiler (ii) Nozzle (iii) Centrifugal pump
(iv) Throttling device (v) Turbine **(10 Marks)**

b. The working fluid in a steady flow process flows at the rate of 220 kg/min. the fluid rejects 100 kJ/s of heat passing through the system. The fluid enters at a velocity of 320 m/s, pressure of 6 bar, internal energy 2000 kJ/kg, specific volume of $0.36\ m^3/kg$ and leaves the system at a velocity of 140 m/s, pressure of 1.2 bar, internal energy 1400 kJ/kg, specific volume of $1.3\ m^3/kg$. Determine the power output in MW. The change in potential energy is neglected. **(10 Marks)**

1 of 2

QUESTION BANK

2 Marks Questions and Answers

Unit I - Basic Concepts and First Law

1. **What is meant by thermodynamic system? How do you classify it?** Thermodynamic system is defined as any space or matter or group of matter where energy transfer or energy conversions are studied. It may be classified into three types: (a) Open system (b) Closed system (c) Isolated system
2. **What is meant by closed system? Give an example.** When a system has only heat and work transfer, but there is no mass transfer, it is called a closed system. **Example:** Piston and cylinder arrangement.
3. **Define open system. Give an example.** When a system has heat, work and mass transfer, it is called an open system. **Example:** Air compressor.
4. **Distinguish between open and closed system.**

S.No	Open System	Closed System
1	Only heat and work will transfer	In addition to heat and work transfer
2	System boundary is fixed	System boundary may or may not change
3	Ex: Piston and cylinder arrangement, Thermal power plant	Ex: Air compressor, boiler

5. **Define isolated system.** Isolated system is not affected by the surroundings. There is no heat, work and mass transfer. In this system total energy remains constant. **Example:** Entire Universe.
6. **Define specific heat capacity at constant pressure.** It is defined as the amount of heat transfer required to raise or lower the temperature of the unit mass of the substance through one degree when the pressure is kept constant. It is denoted by **Cp**.
7. **Define specific heat capacity at constant volume.** It is defined as the amount of heat transfer required to raise or lower the temperature of the unit mass of the substance through one degree when the volume is kept constant. It is denoted by **Cv**.
8. **What is meant by surrounding?** Any other matter outside of the system boundary is called surrounding.
9. **What is boundary?** System and surrounding are separated by an imaginary line called boundary.
10. **What is meant by thermodynamic property?** Thermodynamic property is any characteristic of a substance which is used to identify the state of the system and can be measured, when the system remains in an equilibrium state.
11. **How do you classify the property?** Thermodynamic property can be classified into two types:
12. Intensive or Intrinsic
13. Extensive or Extrinsic property.

14. **Define intensive and extensive properties. Intensive properties:** Properties which are independent of the mass of the system (e.g., Pressure, Temperature, specific volume). **Extensive properties:** Properties whose values depend on the mass of the system (e.g., Total energy, Total volume, weight).
15. **When a system is said to be in "Thermodynamic equilibrium"?** When a system is in thermodynamic equilibrium, it should satisfy the following three conditions: (a) Mechanical Equilibrium – Pressure remains constant. (b) Thermal Equilibrium – Temperature remains constant. (c) Chemical Equilibrium – There is no chemical reaction.
16. **Define Zeroth law and first law of thermodynamics. Zeroth law:** When two systems are separately in thermal equilibrium with a third system, then they themselves are in thermal equilibrium with each other. **First law:** When a system undergoes a cyclic process, net heat transfer is equal to work transfer. $dQ = dW$.
17. **State corollaries of first law of thermodynamics. Corollary I:** There exists a property of a closed system such that a change in its value is equal to the difference between the heat supplied and the work done during any change of state. **Corollary II:** The internal energy of a closed system remains unchanged if the system is isolated from its surroundings. **Corollary III:** A perpetual motion machine of the first kind is impossible.
18. **What is meant by "Perpetual Motion Machine of First Kind"?** PMM of the first kind delivers work continuously without any input. It violates the first law of thermodynamics. It is impossible to construct an engine working with this principle.
19. **What is meant by control volume and control surface? Control Volume:** A fixed region in space chosen for the thermodynamic study of mass balances for flowing systems. **Control Surface:** The boundary of the control volume.
20. **Using Knudsen number define continuum.** The Knudsen number (Kn) is a dimensionless number defined as the ratio of the molecular mean free path length to a representative physical length scale.
21. **Prove that for an isolated system, there is no change in internal energy.** For any isolated system, there is no heat, work and mass transfer. $Q = W = 0$ According to the first law: $Q = W + \Delta U \rightarrow \Delta U = 0$.
22. **What is meant by reversible and irreversible process?** A process is said to be reversible if it traces the same path in the reversed direction. It is possible only when the system passes through a continuous series of equilibrium states. If the system does not pass through continuous equilibrium states, then the process is irreversible.
23. **What is meant by point and path function? Point function (State function):** Properties whose change depends only on the initial and final states (e.g., pressure, temperature, volume, internal energy). **Path function (Process function):** Quantities whose values depend on the path followed (e.g., work and heat).

Unit II - Second Law of Thermodynamics (2 Marks Questions & Answers)

1. State the Kelvin-Planck statement of the second law of Thermodynamics.

It is impossible to construct a heat engine working on a cyclic process, whose only purpose is to convert all the heat energy given to it into an equal amount of work.

2. State the Clausius statement of the second law of Thermodynamics.

It is impossible to construct a device which operates on a cycle and produces no effect other than the transfer of heat from a colder body to a hotter body.

3. Define Heat Engine.

A heat engine is a device which converts heat energy into mechanical work. It operates between a high temperature source and a low temperature sink.

4. Define Thermal Efficiency of a Heat Engine.

Thermal efficiency is the ratio of net work output to the heat input.

$\eta = W_{net} / Q_{in}$

5. What is a Refrigerator?

A refrigerator is a device which removes heat from a low temperature body and rejects it to a high temperature body with the help of external work.

6. Define Coefficient of Performance (COP) of a Refrigerator.

COP = Heat removed from cold body / Work input = Q_c / W

7. Define Heat Pump.

A heat pump is a device which absorbs heat from a low temperature body and delivers it to a high temperature body with the help of external work.

8. Define COP of a Heat Pump.

COP_HP = Heat delivered to hot body / Work input = Q_h / W

9. What is meant by Perpetual Motion Machine of the Second Kind (PMM2)?

It is a device which absorbs heat from a single reservoir and converts it completely into work without rejecting heat to a sink. It violates the second law of thermodynamics.

10. State Carnot's Theorem.

No heat engine operating between two given temperature limits can have higher efficiency than a reversible engine operating between the same temperature limits.

11. What are the corollaries of Carnot's Theorem?

- All reversible engines operating between the same temperature limits have the same efficiency.

- The efficiency of a reversible engine is a function of the temperatures of the source and sink only.

12. Define Thermodynamic Temperature Scale.

It is a temperature scale which is independent of the working substance. Kelvin scale is the thermodynamic temperature scale.

13. What is meant by Entropy?

Entropy is a thermodynamic property which is a measure of the degree of randomness or disorder in a system. It is denoted by 'S'.

14. State the principle of increase of Entropy.

The entropy of an isolated system always increases and remains constant only in a reversible process.

15. Define Availability.

Availability is the maximum useful work that can be obtained from a system when it is brought to the dead state (equilibrium with the surroundings).

16. Define Irreversibility.

Irreversibility is the difference between the available energy and the actual work obtained. It represents the loss of useful work due to irreversibilities.

17. What is meant by Gibbs Free Energy?

Gibbs function (G) = H – TS. It is a measure of the maximum useful work obtainable from a system at constant temperature and pressure.

18. State the Third Law of Thermodynamics (Nernst Law).

The entropy of a pure crystalline substance approaches zero as the absolute temperature approaches zero.

19. What is Clausius Inequality?

For a cyclic process, $\oint (\delta Q / T) \leq 0$. Equality holds for reversible processes and inequality for irreversible processes.

20. Distinguish between Availability and Irreversibility.

Parameter	Availability	Irreversibility
Definition	Maximum useful work	Loss of available energy
Symbol	A	I
Nature	Positive	Always positive

Unit III - Properties of Pure Substances and Steam Power Cycle (2 Marks Questions & Answers)

1. **Define pure substance.** A pure substance is one which has a homogeneous and invariable chemical composition. It may exist in one or more phases.
2. **What is meant by latent heat of fusion?** It is the amount of heat required to convert 1 kg of solid into liquid at constant temperature (melting point).
3. **Define latent heat of vaporization.** It is the amount of heat required to convert 1 kg of liquid into vapour at constant temperature (boiling point).
4. **What is saturation temperature?** It is the temperature at which the liquid and vapour phases coexist in equilibrium at a given pressure.
5. **Define dryness fraction (x).** Dryness fraction is the ratio of the mass of dry steam to the total mass of wet steam. $x = mg / (mf + mg)$ where mg = mass of dry steam, mf = mass of liquid.
6. **Define specific volume of wet steam.** $v = vf + x (vg – vf)$ where vf = specific volume of saturated liquid, vg = specific volume of saturated vapour.
7. **What is superheated steam?** When the temperature of steam is greater than the saturation temperature corresponding to that pressure, it is called superheated steam.
8. **Define enthalpy of wet steam.** $h = hf + x\, hfg$ where hf = enthalpy of saturated liquid, hfg = latent heat of vaporization.
9. **What is meant by critical point?** It is the point at which the saturated liquid and saturated vapour states become identical. Beyond this point, there is no distinction between liquid and vapour phases.
10. **Define triple point.** It is the state at which all the three phases (solid, liquid and vapour) coexist in equilibrium.
11. **What is meant by sensible heat?** It is the heat absorbed by the substance during heating without change of phase.
12. **Draw p-v, T-s and h-s diagrams for steam.** (Describe or sketch in Word: Saturation dome, critical point, wet region, superheated region).
13. **State the importance of Mollier diagram (h-s diagram).** It is very useful in solving problems on turbines and compressors as the vertical distance represents work.
14. **Define steam power cycle.** It is a thermodynamic cycle in which the working fluid is water and steam. The most common is the Rankine cycle.
15. **What are the processes in Rankine cycle?** 1-2: Isentropic compression in pump 2-3: Isobaric heat addition in boiler 3-4: Isentropic expansion in turbine 4-1: Isobaric heat rejection in condenser
16. **Define mean effective pressure.** It is the constant pressure which, if acting on the piston during the power stroke, would produce the same work output as the actual cycle.
17. **What is the effect of superheating on Rankine cycle efficiency?** Superheating increases the average temperature at which heat is added, thereby increasing the efficiency of the cycle.
18. **What is the effect of increasing boiler pressure on Rankine cycle?** Increasing boiler pressure increases the efficiency but also increases the moisture content at turbine exit.
19. **What is reheat cycle?** In reheat cycle, steam is expanded in two stages with reheating in between to reduce moisture content at turbine exit.

20. **What is regenerative cycle?** Bleeding a part of steam from turbine to heat the feed water, thereby increasing efficiency by reducing heat addition in boiler.

Unit IV - Ideal and Real Gases and Thermodynamic Relations (2 Marks Questions & Answers)

1. **Define Ideal Gas.** An ideal gas is a hypothetical gas which obeys the gas laws at all pressures and temperatures. It has no intermolecular forces and the volume occupied by the molecules is negligible.
2. **State the equation of state for an ideal gas.** pV = mRT or pV = nRT where p = pressure, V = volume, m = mass, R = gas constant, T = temperature, n = number of moles.
3. **What is universal gas constant?** It is the gas constant for one mole of gas. Its value is 8.314 kJ/kmol.K.
4. **Define characteristic gas constant.** It is the gas constant per unit mass of the gas. R = Ru / M (where M = molecular weight).
5. **State Avogadro's law.** Equal volumes of all gases at the same temperature and pressure contain equal number of molecules.
6. **Define compressibility factor.** Compressibility factor Z = (pV) / (mRT) for real gases. For ideal gases Z = 1.
7. **What is meant by real gas?** A real gas deviates from ideal gas behaviour at high pressures and low temperatures.
8. **State Van der Waals equation of state.** $(p + a/V^2)(V - b) = RT$ where a and b are Van der Waals constants.
9. **What is meant by reduced properties?** Reduced pressure Pr = p / pc Reduced temperature Tr = T / Tc Reduced volume Vr = V / Vc (where pc, Tc, Vc are critical pressure, temperature and volume).
10. **State the law of corresponding states.** All gases when compared at the same reduced temperature and reduced pressure have approximately the same compressibility factor.
11. **Define Joule-Thomson coefficient.** It is the rate of change of temperature with respect to pressure at constant enthalpy. $\mu = (\partial T/\partial p)h$
12. **What is inversion temperature?** It is the temperature at which the Joule-Thomson coefficient changes sign ($\mu = 0$).
13. **Define Maxwell's relations.** They are a set of thermodynamic relations derived from the exact differentials of thermodynamic potentials.
14. **Write any two Maxwell's relations.** $(\partial T/\partial V)s = -(\partial p/\partial S)V$ $(\partial T/\partial p)s = (\partial V/\partial S)p$
15. **Define Helmholtz free energy.** Helmholtz function F = U – TS. It represents the maximum useful work obtainable from a system at constant temperature and volume.
16. **Define Gibbs free energy.** Gibbs function G = H – TS. It represents the maximum useful work obtainable from a system at constant temperature and pressure.
17. **What is the Clausius-Clapeyron equation used for?** It is used to determine the variation of saturation pressure with temperature during phase change.
18. **State the conditions for a process to be reversible.** The process must be quasi-static and there should be no dissipative effects like friction, unrestrained expansion, etc.
19. **Define fugacity.** Fugacity is a property which represents the effective pressure of a real gas. For ideal gases, fugacity equals pressure.
20. **What are the uses of thermodynamic relations?** They are used to calculate properties which cannot be measured directly, such as entropy, internal energy, etc., from measurable properties like pressure, volume and temperature.

Unit V - Gas Mixtures and Psychrometry (2 Marks Questions & Answers)

1. **Define gas mixture.** A gas mixture is a mixture of two or more gases which do not chemically react with each other.
2. **State Dalton's law of partial pressures.** The total pressure of a mixture of gases is equal to the sum of the partial pressures of the individual gases. $P = p_1 + p_2 + p_3 + \ldots + p_n$
3. **State Amagat's law (or Leduc's law).** The total volume of a mixture of gases is equal to the sum of the partial volumes of the individual gases at the same temperature and pressure.
4. **Define partial pressure.** It is the pressure which each component of the gas mixture would exert if it alone occupied the entire volume of the mixture at the same temperature.
5. **Define mole fraction.** Mole fraction of a component is the ratio of the number of moles of that component to the total number of moles in the mixture.
6. **Define mass fraction.** Mass fraction of a component is the ratio of the mass of that component to the total mass of the mixture.
7. **What is meant by apparent molecular weight of a gas mixture?** It is the molecular weight of the mixture calculated as M = m_total / n_total.
8. **Define psychrometry.** Psychrometry is the study of the properties of atmospheric air and water vapour mixtures.
9. **Define dry bulb temperature (DBT).** It is the temperature of air measured by an ordinary thermometer.
10. **Define wet bulb temperature (WBT).** It is the temperature of air measured by a thermometer whose bulb is covered with a wet wick and exposed to air flow.
11. **Define dew point temperature (DPT).** It is the temperature at which the water vapour present in the air begins to condense when cooled at constant pressure.
12. **Define relative humidity (φ).** Relative humidity is the ratio of the mass of water vapour in a given volume of moist air to the mass of water vapour in the same volume of saturated air at the same temperature and pressure. $\varphi = (pv / pg) \times 100\%$ where pv = partial pressure of vapour, pg = saturation pressure.
13. **Define specific humidity (ω) or humidity ratio.** It is the mass of water vapour present per unit mass of dry air. $\omega = 0.622 \times (pv / (pa))$ where pa = partial pressure of dry air.
14. **Define degree of saturation.** It is the ratio of the specific humidity of moist air to the specific humidity of saturated air at the same temperature.
15. **What is a psychrometric chart?** It is a graphical representation of the various thermodynamic properties of moist air. It is very useful for solving air conditioning problems.
16. **What are the important psychrometric processes?**

- Sensible heating
- Sensible cooling
- Humidification
- Dehumidification
- Cooling and dehumidification
- Heating and humidification
- Adiabatic mixing of air streams

17. **Define bypass factor.** It is the ratio of the difference between the leaving air temperature and the coil temperature to the difference between the entering air temperature and the coil temperature.
18. **What is meant by sensible heat factor (SHF)?** Sensible heat factor is the ratio of sensible heat to the total heat (sensible + latent heat).
19. **Define apparatus dew point (ADP).** It is the effective surface temperature of the cooling coil.
20. **What is the use of psychrometric chart in air conditioning?** It is used to determine the properties of air at various states and to calculate the heating/cooling loads, moisture removal, etc.

Module IV - Reciprocating Compressors

4.1 Introduction Compressors are devices used to increase the pressure of a gas by reducing its volume. They are widely used in refrigeration, air conditioning, and pneumatic systems.

4.2 Uses of Compressed Air

- Operating pneumatic tools
- Starting of IC engines
- Spray painting
- Air brakes in vehicles
- Cleaning and inflating tyres
- Refrigeration and air conditioning systems

4.3 Classification of Compressors

1. **Positive Displacement Compressors**
   - Reciprocating compressors
   - Rotary compressors (vane, screw, roots blower)
2. **Dynamic Compressors** (Steady Flow)
   - Centrifugal compressors
   - Axial flow compressors

4.4 Single Stage Reciprocating Air Compressor It consists of a cylinder, piston, inlet and outlet valves. The piston moves to and fro inside the cylinder to compress the air.

4.5 Compression Processes (a) **Isentropic (or Adiabatic) Compression** pV^γ = Constant Work done = $[\gamma/(\gamma-1)]\ mRT_1\ [(p_2/p_1)^{((\gamma-1)/\gamma)} - 1]$

(b) **Polytropic Compression** pV^n = Constant Work done = $[n/(n-1)]\ mRT_1\ [(p_2/p_1)^{((n-1)/n)} - 1]$

(c) **Isothermal Compression** pV = Constant Work done = $mRT\ \ln(p_2/p_1)$

4.6 Power Required for Driving the Compressor Power = Work done per cycle × (N/60) where N = speed in rpm.

4.7 Effect of Clearance Volume Clearance volume is the volume remaining in the cylinder when the piston is at the end of the delivery stroke. It reduces the volumetric efficiency of the compressor.

Volumetric Efficiency (η_{vol}) $\eta_{vol} = 1 + C - C\ (p_2/p_1)^{(1/n)}$ where C = Clearance ratio = Vc / Vs

4.13 Multi-Stage Air Compressor Used to reduce work input and to limit the maximum temperature. Air is compressed in stages with intercooling between stages.

Advantages of Multi-Stage Compression

- Reduced work input
- Better volumetric efficiency
- Reduced leakage losses
- Smaller flywheel size
- Reduced delivery temperature

4.14 Intercoolers Intercoolers are heat exchangers used to cool the air between stages of compression. Perfect intercooling brings the air back to the initial temperature after each stage.

HEAT ENGINES - Module I: Introduction

Comparison between External Combustion Engine and Internal Combustion Engine

S.No	External Combustion Engine	Internal Combustion Engine
1	Combustion takes place outside the cylinder	Combustion takes place inside the cylinder
2	Working fluid does not mix with products of combustion	Working fluid mixes with products of combustion
3	Examples: Steam engines, Steam turbines	Examples: Petrol, Diesel, Gas engines
4	Lower thermal efficiency	Higher thermal efficiency
5	Heavy and bulky	Compact and lighter

Main Components of Reciprocating IC Engines

- Cylinder
- Piston
- Connecting rod
- Crankshaft
- Crankcase
- Cylinder head
- Valves (Inlet & Exhaust)
- Spark plug / Fuel injector

Terminology used in IC Engines

- Bore (D)
- Stroke (L)
- Swept volume / Piston displacement
- Clearance volume
- Compression ratio ($r = (Vs + Vc)/Vc$)
- Mean effective pressure (MEP)
- Indicated power (IP)
- Brake power (BP)
- Mechanical efficiency = BP / IP

Four Stroke Engine One cycle is completed in four strokes of the piston (two revolutions of the crankshaft).

1. Suction stroke
2. Compression stroke
3. Expansion / Power stroke
4. Exhaust stroke

Two Stroke Engine One cycle is completed in two strokes of the piston (one revolution of the crankshaft). Power stroke occurs in every revolution.

Comparison of Four Stroke and Two Stroke Engine

Parameter	Four Stroke	Two Stroke
No. of power strokes	One per 2 revolutions	One per revolution
Power output	Less	More (for same size)
Thermal efficiency	Higher	Lower
Cooling & Lubrication	Easier	Difficult
Applications	Cars, trucks	Motorcycles, scooters

Comparison of SI and CI Engines

Parameter	SI Engine (Petrol)	CI Engine (Diesel)
Fuel	Petrol	Diesel
Ignition	Spark ignition	Compression ignition
Compression ratio	6 to 12	14 to 22
Air-Fuel ratio	10:1 to 20:1	18:1 to 70:1
Efficiency	Lower	Higher

Valve Timing Diagram

- Theoretical valve timing diagram
- Actual valve timing diagram (considering valve overlap, lead and lag)

Port Timing Diagram (for 2-Stroke Engines)

FUELS & FUEL INJECTION

Classification of Petroleum Fuels used for IC Engines

1. Liquid fuels
   - Petrol (Gasoline)
   - Diesel
   - Kerosene
2. Gaseous fuels
   - LPG, CNG, Hydrogen, Producer gas, etc.

Refinery Processes

- Fractional distillation
- Cracking (Thermal & Catalytic)
- Reforming
- Polymerization, etc.

Alternative Fuels

- Methanol
- Ethanol
- Hydrogen
- Natural Gas (CNG)
- Producer Gas
- Blast Furnace Gas
- Coke Oven Gas

General Fuel Specifications

- Relative density (Specific gravity)
- Fuel composition
- Specific heating value (Calorific value)
- Flash point
- Viscosity
- Surface tension
- Freezing point

Important Fuel Specifications for Gasoline

(a) Gasoline Volatility – Measured by distillation curve

(b) Antiknock Quality – Measured by Octane Number

Important Fuel Specifications for Diesel

- Viscosity
- Cetane Number (Measure of ignition quality)
- Flash point

Carburetion

- Mixture requirements for steady state and transient operation
- Simple Carburettor – Main metering system, Idling system, Power enrichment system, Acceleration pump, Choke

Carburettor Types
- Solex Carburettor
- Carter Carburettor
- S.U. Carburettor

Drawbacks of Modern Carburettor
Petrol Injection Systems:
- Continuous Injection
- Timed Injection System
- Electronic Fuel Injection (L-Jetronic, etc.)

Diesel Injection System
- Air Injection System
- Individual Pump and Injector (Jerk Pump)
- Common Rail System
- Distributor System
- Types of Nozzles
- Combustion Chambers: Open, Pre-combustion

Module III - IGNITION SYSTEM

Battery or Coil Ignition System
- Circuit diagram includes battery, ignition switch, coil, distributor, spark plug, condenser.

Magneto Ignition System
- Used in two-wheelers and small engines.

Advantages of Electronic Ignition System
- Better spark, longer life, precise timing.

Firing Order
- Sequence in which cylinders receive spark (e.g., 1-3-4-2 for 4-cylinder engine).

Ignition Timing & Ignition Advance
- Spark occurs before TDC to account for flame propagation time.

Stages of SI Engine Combustion
I. Ignition lag
II. Flame propagation
III. Afterburning

Detonation / Knocking in SI Engines
- Causes, effects, and methods to reduce knocking (higher octane fuel, better chamber design, etc.).

Pre-ignition
SI Engine Combustion Chamber Designs

Stages of Combustion in CI Engines
- Ignition delay period
- Uncontrolled combustion
- Controlled combustion

Diesel Knock
- Variables affecting delay period: Injection timing, load, intake temperature, engine speed, combustion chamber design, swirl rate, oxygen concentration.

FUELS & FUEL INJECTION

Classification of Petroleum Fuels used for IC Engines

1. Liquid fuels
   - Petrol (Gasoline)
   - Diesel
   - Kerosene
2. Gaseous fuels
   - LPG, CNG, Hydrogen, Producer gas, etc.

Refinery Processes

- Fractional distillation
- Cracking (Thermal & Catalytic)
- Reforming
- Polymerization, etc.

Alternative Fuels

- Methanol
- Ethanol
- Hydrogen
- Natural Gas (CNG)
- Producer Gas
- Blast Furnace Gas

- Coke Oven Gas

General Fuel Specifications

- Relative density (Specific gravity)

- Fuel composition

- Specific heating value (Calorific value)

- Flash point

- Viscosity

- Surface tension

- Freezing point

Important Fuel Specifications for Gasoline

(a) Gasoline Volatility – Measured by distillation curve

(b) Antiknock Quality – Measured by Octane Number

Important Fuel Specifications for Diesel

- Viscosity

- Cetane Number (Measure of ignition quality)

- Flash point

Carburetion

- Mixture requirements for steady state and transient operation

- Simple Carburettor – Main metering system, Idling system, Power enrichment system, Acceleration pump, Choke

Carburettor Types

- Solex Carburettor
- Carter Carburettor
- S.U. Carburettor

Drawbacks of Modern Carburettor

Petrol Injection Systems:

- Continuous Injection
- Timed Injection System
- Electronic Fuel Injection (L-Jetronic, etc.)

Diesel Injection System

- Air Injection System
- Individual Pump and Injector (Jerk Pump)
- Common Rail System
- Distributor System
- Types of Nozzles
- Combustion Chambers: Open, Pre-combustion

Module III - IGNITION SYSTEM

Battery or Coil Ignition System

- Circuit diagram includes battery, ignition switch, coil, distributor, spark plug, condenser.

Magneto Ignition System

- Used in two-wheelers and small engines.

Advantages of Electronic Ignition System

- Better spark, longer life, precise timing.

Firing Order

- Sequence in which cylinders receive spark (e.g., 1-3-4-2 for 4-cylinder engine).

Ignition Timing & Ignition Advance

- Spark occurs before TDC to account for flame propagation time.

Stages of SI Engine Combustion

I. Ignition lag

II. Flame propagation

III. Afterburning

Detonation / Knocking in SI Engines

- Causes, effects, and methods to reduce knocking (higher octane fuel, better chamber design, etc.).

Pre-ignition

SI Engine Combustion Chamber Designs

Stages of Combustion in CI Engines

- Ignition delay period

- Uncontrolled combustion

- Controlled combustion

Diesel Knock

- Variables affecting delay period: Injection timing, load, intake temperature, engine speed, combustion chamber design, swirl rate, oxygen concentration.

FLUID MECHANICS AND MACHINERY - TWO MARK QUESTIONS

1. **Define density or mass density.** Density (ρ) = Mass / Volume (kg/m³). $\rho_{water} = 1000$ kg/m³.
2. **Define specific weight.** Specific weight (γ) = Weight / Volume (N/m³).
3. **Define dynamic viscosity.** It is the ratio of shear stress to the rate of shear strain. 1 N-s/m² = 1 Pa-s = 10 Poise.
4. **Define compressibility.** It is the reciprocal of bulk modulus of elasticity.
5. **Define surface tension.** Surface tension (σ) = Force per unit length (N/m). $p_{total} = p_{inside} + p_{atm}$ ($p_{atm} = 101.325 \times 10^3$ N/m²).
6. **Define control volume.** It is a fixed region in space through which fluid flows.
7. **Write the continuity equation.** For steady flow: $\rho_1 A_1 V_1 = \rho_2 A_2 V_2$.
8. **List the types of fluid flow.**

- Steady and unsteady flow
- Uniform and non-uniform flow
- Laminar and turbulent flow
- Compressible and incompressible flow
- Rotational and irrotational flow

9. **Define steady and unsteady flow.** Steady flow: $\partial V/\partial t = 0$, $\partial p/\partial t = 0$, $\partial \rho/\partial t = 0$. Unsteady flow: Any of the above is non-zero.
10. **Define compressible and incompressible flow.** Compressible flow: $\rho \neq$ constant. Incompressible flow: ρ = constant.
11. **Define irrotational flow.** Flow in which vorticity is zero.
12. **Write Bernoulli's equation applied between two sections.** $p/\rho g + V^2/2g + z$ = constant.
13. **State the assumptions used in deriving Bernoulli's equation.**

- Steady flow
- Inviscid flow (no friction)
- Incompressible flow
- Flow along a streamline

14. **Define Impulse-Momentum Equation.** $F\,dt = d(mV)$
15. **Mention the range of Reynolds number for laminar and turbulent flow in a pipe.** Laminar: $Re < 2000$ Turbulent: $Re > 4000$ Transition: $2000 < Re < 4000$
16. **What does Hagen-Poiseuille's equation refer to?** It gives pressure drop in laminar flow through a circular pipe. $(p_1 - p_2)/\rho g = h_f = 32\mu UL / (\rho g D^2)$
17. **Write the expression for shear stress.** $\tau = -(\partial p/\partial x)(r/2)$ $\tau_{max} = -(\partial p/\partial x)(R/2)$
18. **Define kinetic energy correction factor.** It accounts for non-uniform velocity distribution.
19. **Define boundary layer.** It is the thin layer of fluid near the surface where viscous effects are significant.
20. **Define displacement thickness, momentum thickness, and energy thickness.** $\delta^* = \int[1 - (u/U)]\,dy$ $\theta = \int[(u/U) - (u/U)^2]\,dy$ $\delta^{**} = \int[(u/U) - (u/U)^3]\,dy$

21. **Explain major and minor losses in a pipe.** Major losses: Due to friction (Darcy-Weisbach equation). Minor losses: Due to sudden expansion, contraction, bends, valves, etc.
22. **State Darcy-Weisbach equation.** $h_f = f (L/D) (V^2/2g)$
23. **Define dimensional analysis.** It is a method of reducing the number of variables in a physical problem by forming dimensionless groups.
24. **State Buckingham's π theorem.** If there are n variables with k fundamental dimensions, then there are (n - k) dimensionless π terms.
25. **Define model and prototype.** Model is the scaled version; prototype is the actual machine.
26. **Define geometric, kinematic, and dynamic similarities.**
27. **State Reynolds model law, Froude's model law, Euler's model law.**
28. **Define pump and centrifugal pump.** Pump is a device which adds energy to a fluid. Centrifugal pump uses centrifugal force.
29. **Define specific speed of a centrifugal pump.**
30. **Define slip and % slip in reciprocating pump.** Slip = $Q_{theoretical} - Q_{actual}$
31. **Define cavitation.** It is the formation and collapse of vapour bubbles causing damage.
32. **Define draft tube and its uses.** Draft tube recovers kinetic energy at the outlet of reaction turbines.
33. **Define specific speed of a turbine.**
34. **Define hydraulic efficiency, mechanical efficiency, volumetric efficiency, and overall efficiency.**

www.ingramcontent.com/pod-product-compliance
Lightning Source LLC
LaVergne TN
LVHW060034210926
848481LV00013B/3863
9798223617501